D1505062

A HOT MESS

A HOT

MESS

How the Climate Crisis
Is Changing Our World

JEFF FLEISCHER

ZEST BOOKS
MINNEAPOLIS

Special thanks to Dr. Carlos E. Del Castillo, Oceanographer, for reviewing the text.

Text copyright © 2021 by Jeff Fleischer

Zest Books™
An imprint of Lerner Publishing Group, Inc.
241 First Avenue North
Minneapolis, MN 55401 USA

For reading levels and more information, look up this title at www.lernerbooks.com.
Visit us at zestbooks.net 🖪 🖪

Main body text set in Minion Pro
Typeface provided by Adobe Systems.

Library of Congress Cataloging-in-Publication Data

Names: Fleischer, Jeff, author.
Title: A hot mess : how the climate crisis is changing our world / Jeff Fleischer.
Description: Minneapolis : Zest Books, [2021] | Includes bibliographical references. |
  Audience: Ages 11–18 | Audience: Grades 7–9 | Summary: "Drawing on real-life
  situations and stories, journalist Jeff Fleischer takes an informed, approachable
  look at how our world will change as a result of the climate crisis, addressing
  sea levels, extreme weather, drought, extinction, and migration"— Provided by
  publisher.
Identifiers: LCCN 2020028748 (print) | LCCN 2020028749 (ebook) |
  ISBN 9781541597761 (library binding) | ISBN 9781541597778 (paperback) |
  ISBN 9781728419145 (ebook)
Subjects: LCSH: Climatic changes—Juvenile literature. | Global temperature
  changes—Juvenile literature.
Classification: LCC QC903.15 .F54 2021 (print) | LCC QC903.15 (ebook) | DDC
  363.738/74—dc23

LC record available at https://lccn.loc.gov/2020028748
LC ebook record available at https://lccn.loc.gov/2020028749

Manufactured in the United States of America
3-52529-48367-2/2/2022

33614082950022

CONTENTS

INTRODUCTION

I WAS ONLY about a year old when the British colony called the Ellice Islands became the independent nation of Tuvalu. Other than where it was on my globe, what the flag looked like, and a few other basic facts, I didn't know much about Tuvalu for the first twenty-six years of my life. While working at a newspaper in Australia after graduate school, I learned more about Tuvalu. The small country was already dealing with massive annual flooding, and its leaders had begun planning for worst-case scenarios.

By that time, the odds were already solid that, if I made it to my life expectancy, I was probably going to outlast Tuvalu. Some scientific predictions made in 2007 gave the country only about thirty or forty years before it would become uninhabitable. Tuvaluans knew it would take a massive international effort to save nations like theirs from vanishing beneath the waves because of climate change.

I spent most of 2008 abroad on a journalism grant, learning more about Tuvalu and how people there were dealing with the changing climate—both what they were experiencing at the time and how they were preparing for the future. That included more than a month spent conducting interviews on Funafuti, the country's main island and capital. Funafuti is a coral atoll (see page 90) that is home to about half of the country's eleven thousand or so people.

There were a few places on Funafuti's main islet where the land was so narrow that I could look straight ahead and see both the dark-blue ocean that surrounds the ring-shaped atoll and the aquamarine freshwater lagoon in the middle of it. Elsewhere, on humid days after a warm sun shower, rainwater still pooled in spots on the ground, draining slowly. While visiting the office of the country's meteorological service, I saw a wall decorated with photos of the staff members during the aftermath of annual high tides, posing in a line with water covering much of their legs.

Tuvalu is one of several Pacific island nations experiencing more flooding and loss of land due to rising sea levels.

Along the coastline, I found several places where recent erosion was obvious and where the ocean tide had taken land once passed down within families for generations. It was easy to see what a small margin of error Tuvalu had when it came to dealing with flooding from rising seas. "I worry about when we have to leave," one young police officer told me. "Maybe it's still here for my children, but what about my grandchildren? Or their children?"

Really think about that for a moment. We all have some uncertainty about our own futures—a million things can always change, for better or worse—but most of us expect our society to still be there as we get older and to stay in many ways the same. What if that weren't true? What would you do?

As Tuvaluans talked to me about the future, several worried about whether their culture and language would even exist in a few generations if the population of their small country needed to disperse among several other places. (Tuvalu spreads over an area of ocean larger than France, but its total land is only about one-tenth the size of Washington, DC.)

Most of the people I interviewed felt torn between trying to plan a life in their homeland and preparing for the possibility that they might need to find a new one. That was especially true for people in their twenties and thirties, who had grown up in a place with a tight-knit community but without much of a formal economy. I talked with Tuvaluans in New Zealand who had signed up to pick apples and kiwifruit there for a few weeks as seasonal work. Some were just trying to earn extra money for their families, but several talked about it as a way to prepare for an eventual move.

People I met on Funafuti were, in increasing numbers, studying at universities in Australia, New Zealand, or Fiji, while others took part in guest-worker programs to develop job skills they might need down the line and to build up savings they might require one day for migration. Many Tuvaluans have been working hard to help their islands adapt to the challenges of climate change, and the country's leaders have been an important presence at international negotiations, urging polluting nations to reduce emissions and switch to greener technology.

"We have to look at the next ten or twenty years and think, 'What would we be like?'" the then general manager of Tuvalu's philatelic bureau (stamps from Tuvalu are a big deal with international collectors) told me. "Do we still have enough to survive on an island like this?" We don't know for sure when the answer to that question will become "no."

I chose Tuvalu for that research project not only because it was a fascinating story, but because the changes it was experiencing served as a real-time warning for the whole planet. There were others, of course: already-severe droughts in China, for example, or once-vast glaciers in North America that have shrunk or disappeared. But this was an entire country under serious stress.

At that point in time, people in my home country, the United States, often thought about climate change as a future problem—if they thought about it at all—because the effects they were already seeing weren't yet crises. In plenty of places, climate change was easy enough to ignore. People who either didn't know any better or didn't want to admit the seriousness of the situation saw it as someone else's problem, part of a normal pattern, or "not that bad." While I was still abroad in 2008, what became known as the Great Recession began, and Americans' attention understandably turned to the more immediate economic crisis. Still, it's not as if climate change stopped being a problem during that time or stopped getting worse.

More than a decade later, the situation is no longer easy for those same people to ignore. There's still a tendency to talk about climate change in terms of the yet-to-come, worst-case forecasts, but it's been with us for a while now—and we're running short on time to stop those worst cases from happening.

With that in mind, *A Hot Mess* will explain the basic science behind climate change. Not just what's happening and what's causing it, but some of the mechanics of how things work. It will answer questions such as why carbon emissions heat the planet, why the sea level is rising, and why certain places are getting colder even though the planet overall is growing hotter.

But this isn't exclusively a science book. Lots of books focus only on the science, and we learn more about climate science all the time thanks to new data and studies. Instead, *A Hot Mess* is also designed to explain what the changing climate means for people and the ways it's affecting human (and other) life on Earth. For that reason, each chapter in this book focuses on an aspect of climate change, including examples from around the globe of what's already happening and what we can expect in the near future. Different regions face different threats, but nowhere on Earth is unaffected.

One of the challenges of writing any book on this subject is how quickly the pace of climate change has picked up in a fairly short amount of time. As I worked on the book, there was a new example of the severity of climate change—or an ongoing example getting worse—nearly every day. During that time, some scientists started to warn the public that we might have already passed the long-feared "tipping point," after which it would be too late to stop some of the worst aspects of climate change. As the title of a recent *Verge* article aptly said of the period from 2009 to 2019, "This Was the Decade Climate Change Slapped Us in the Face."

Nobody needs to travel far from home, much less to the other side of the planet, to see climate change in action. Even where I live in Chicago, Illinois—far from the coasts and in a generally temperate area—it's impossible to miss. Just about a mile from my apartment on the city's north side, there's a beach along Lake Michigan known as Juneway Beach. Or, more accurately, there *was* a beach. What used to be a sandy shore is now covered by lake water, and the Army Corps of Engineers built a concrete wall a few years back to keep the rising water from encroaching on the shoreline even more.

On one winter Saturday while I was writing in that same apartment, huge waves crashed over the shoreline all along Chicago's famous Lake Shore Drive, causing widespread flooding in several neighborhoods. After that weekend's storms, city officials announced two more beaches on the far north side, Rogers Beach and Howard Beach, were too far gone to save. The city would instead build more walls to try to limit erosion. As the local alderman put it, "The lake has spoken."

During the same weekend, Australia was dealing with an extended drought that turned into months of deadly bushfires that burned across much of the country and put some of the world's most amazing wildlife at risk of extinction. (We'll talk a lot more about that situation later in the book.) It was such a bad period for the planet that the London-based *Guardian* newspaper interviewed a range of scientists about how they cope with the "ecological grief" that comes from seeing what's happening to the natural world they've devoted their lives to studying. One Canada-based scientist described it as "a slow and cumulative grief without end."

The teens and young adults of the twenty-first century are probably going to be more affected by climate change than any other generation in human history. As we'll discuss in the first chapter, what we're living with now is more than a century in the making and a problem that's gathered momentum like a snowball rolling downhill. (That's an analogy fewer people will be able to picture in the near future.)

That's also why, since the early 1990s, most of the world's countries have taken part in negotiations—and eventually agreements—to try to reduce global emissions and decrease the amount of carbon in the atmosphere. However, the goals haven't been aggressive enough, and too many countries have failed to meet them anyway. Worldwide action is going to be needed, and we'll talk more about that later in the book.

Members of 350.org in Denmark stand in solidarity with the people of Tuvalu. For more information on 350.org and other climate activist organizations, see page 183.

It will take a lot more effort and money to undo the damage from the impacts of climate change than it would to prevent those same impacts from happening in the first place. That's why the decades-long efforts of people trying to deny climate change, whether they're just online trolls or oil and coal companies spreading misinformation, have been so dangerous. The more delays they cause, the harder they make it to fix the problem or prevent certain effects. Some of those types have since pivoted to arguing that it's already too late to do anything about the climate crisis and we should just give up. They're wrong about that too.

On a typically humid afternoon during my visit to Tuvalu, I was speaking with the country's assistant environmental minister about a range of subjects related to climate change. I asked him if he found that most people understood what was happening.

He nodded, but explained that even there, on the front lines of climate change, he had to deal with some skepticism and inaction. "There are people who think that everything will be fine," he said, "because the Bible tells them that God promised Noah he will not flood the world again." I told him I had found a few examples of that exact explanation in my short time on Funafuti. While people generally seemed very worried about the rising tide, a few insisted it wouldn't happen and cited the Noah's ark story as proof.

When I asked him what the government could do to educate people in situations like that, he leaned forward and said, "I tell them that God is not the one bringing the floods now. People are doing it, and people need to be the ones to stop it."

That seems as good a place to begin as any.

CHAPTER 1
CLIMATE CHANGE 101

There's one issue that will define the contours of this century more dramatically than any other, and that is the urgent threat of a changing climate.

—then US president Barack Obama, speaking at the World Economic Forum in 2015

BEFORE WE GET into the impacts of climate change and all the ways they're already making life on Earth more difficult, we'll first discuss what climate change is, how it works, and why humankind is making it a lot worse.

Depending on how good your science classes were (and how well you remember them), some of what's included in this chapter might be more of a refresher than new information. But the unfortunate reality is that not everyone has had access to good information. Science denial is big business. Companies that benefit from it have spent many years and many millions of dollars trying to spread skepticism, and social media algorithms have made it easy for bad actors to pass along lies and misleading information. (Don't worry; we'll debunk a lot of that throughout this book.) While this chapter will be brief, it's worth providing a quick overview of the science. Climate change is real. Humans are causing most of it and are making it substantially worse as time goes on. And even in a best-case scenario, it's going to dramatically change the way we live.

If summers feel warmer to you now than they did when you were a little kid, that's not your memory playing tricks on you. The twenty hottest years on record in human history all happened after 1998. The ten hottest have all come since 2005 and the top seven since 2014.

That research comes directly from the National Aeronautics and Space Administration (NASA) and the National Oceanic and Atmospheric Administration (NOAA), the US government agencies that track climate change. The reality is that since people started studying the annual global temperature in the late 1800s, it's gone up dramatically. And in 2018, the United Nations Intergovernmental Panel on Climate Change (IPCC) warned that the worst impacts of climate change will definitely occur if we allow the planet to get more than 2.7°F (1.5°C) hotter than it was before the Industrial Revolution started in the 1700s. We're already past 1.8°F (1°C) hotter, or about two-thirds of the way there. Don't forget; that warning is about preventing the most catastrophic outcomes. Plenty of awful things have already happened, and more will happen even if we manage to avoid the catastrophe level.

The summer of 2019 brought unprecedented temperatures, with June and then July setting records for the hottest month yet. A heatwave in Europe that summer broke national records in at least five countries, killed more than twenty-five hundred people and thousands of farm animals, and overheated much of the

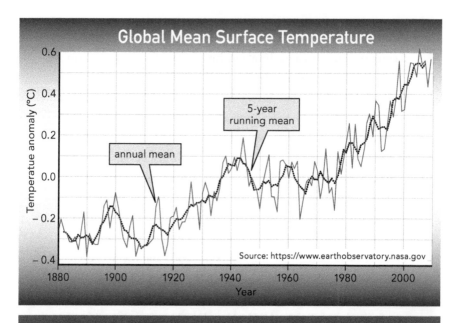

Global Mean Surface Temperature

5-year running mean

annual mean

Source: https://www.earthobservatory.nasa.gov

Since the start of the Industrial Revolution, scientists have kept track of the global mean (average) surface temperature. A clear trend emerged from the data: the planet is getting warmer.

Northern Hemisphere. And we are on track to keep breaking those kinds of records. Even if a lot of things improve, climate change isn't a problem that's going away anytime soon, so it's one worth understanding.

What Are We Talking About?

To best understand climate change, you should understand some of the key terms scientists and other experts use.

First, although they're related, climate and weather are not the same thing. Weather is what the atmosphere is like at a specific time in a specific place. It's temporary and local. As many people in the northern United States can attest, weather can change a few times in the same day—hail can fall on a warm summer day, or rain can quickly turn to snow if the temperature drops over the course of a few hours. In a desert like the Sahara in northern Africa, the average temperature can go from about 100°F (38°C) in the daytime to 25°F (−4°C) the same night.

Climate, on the other hand, is what the atmosphere is like over the long term. When you describe what the weather's *normally* like somewhere, you're really talking about the climate. For example, Florida has a subtropical climate. The weather is hot there most of the time and humid during the rainy season. That doesn't mean it's hot every single day of the year; weather can change there just like it can anywhere else. Southern California has a dry climate, but rain still falls sometimes. Western Washington state has a wet climate, but has its share of sunny days. There might be rain in one town and not in the one right next to it, but both towns probably have a similar climate.

That might seem like a small point, but it's an important one. A warming climate doesn't mean snow will never occur in colder places or that every day will be hot. Instead, the planet, as a whole, will be hotter on an average day than it used to be. This is another key distinction. When we talk about climate change, we're talking about *global* climate change. Scientists discussing numbers related to climate change are usually talking about the climate of Earth in general, rather than a specific part of the planet.

Speaking of global climate change, you may have heard the term *global warming* before. When scientists in the 1980s and 1990s talked about the amount of greenhouse gases in the atmosphere raising Earth's temperature, they usually used that term to describe it. "Global warming" is still accurate—after all, the globe is getting warmer on average. (Richard Betts, a prominent government scientist in the United Kingdom [UK], has suggested using "global heating" as a more accurate term, so some publications in the UK and other countries use that instead; it describes the exact same phenomenon.)

Over time, however, it became clear that the temperature wasn't the only aspect of the climate that's changing—the planet getting warmer has had far-reaching consequences, including extreme weather events, droughts, and melting sea ice. Winters can get colder in certain places, and other areas might get more rainfall or snowfall than they used to get, so the warming world doesn't always *feel* like it's getting warmer everywhere. The term *climate change* captures these consequences in addition to the global temperature increases. That's part of the reason it has become the more popular term. Neither is wrong, but global warming is just one part of climate change. And as the impacts of that climate change become increasingly severe, the term *climate crisis* has become a popular way to describe the threat the world faces.

How Climate Change Works: A Quick Science Lesson

Earth's temperature has a lot to do with its atmosphere (the layer of gases that surrounds the planet) and the components that form it. The change in those components—particularly in the amounts of carbon and methane in the atmosphere—is what's changing our climate.

Of course, radiation from the sun is the main source of light and heat for the planet. Because of how Earth tilts on its axis, areas around the equator—halfway between the north and south poles— are closest to the sun, get the most direct sunlight, and are the warmest parts of the planet. As you move farther north or south, the temperature usually gets colder because the Arctic Circle and Antarctica receive the least direct sunlight and, because of that, are the chilliest places on Earth.

Earth absorbs about 70 percent of the heat it gets from the sun, and it reflects the other 30 percent back into space. That's a natural, normal process—it helps keep the planet at a habitable temperature, allowing life on Earth to exist. Everything from the oceans and soil to plants and animals absorbs heat from the sun. Clouds, snow, and ice are among the things that reflect the sun's heat back, so having them around is crucial for keeping the planet from getting too hot.

Here's where the atmosphere comes into play. The vast majority of it—about 99 percent—is made up of just two elements, nitrogen and oxygen. The last 1 percent consists of lots of other gases, including carbon dioxide, methane, nitrous oxide, halocarbons, and water vapor. These are called greenhouse gases because they produce what's known as the *greenhouse effect.*

Even though they're just a small part of the atmosphere, greenhouse gases have a big impact. When Earth's surface reflects some of the sun's energy back out, the greenhouse gases in the atmosphere block part of that heat from escaping. That keeps the planet warmer. It's not a perfect analogy, but the process is a bit like how the glass walls of a greenhouse trap heat, which helps the plants inside grow, even in cold weather (and that's what gives the greenhouse effect its name).

Greenhouse gases aren't bad in and of themselves. They're supposed to be in the atmosphere. They're one of the reasons why Earth is warm enough to support the kinds of plant and animal life that live here, including us. Without greenhouse gases, the average surface temperature of Earth would be almost 60°F (15.6°C) colder.

The problem is *how much* of those gases is in the atmosphere. The amount of carbon dioxide there now is higher than it's been in at least the last eight hundred thousand years. In other words, it's higher now than it's been at any point in the time that *Homo sapiens* have existed. Humans like us have only been around for about 40 percent of that time.

Earth is more than 4.5 billion years old, and its climate has changed many times in that span. When the planet was new and for millions of years after that, it was incredibly hot. But over time,

A GREENHOUSE OUT OF CONTROL

To get an idea of how greenhouse gases can make a planet unlivable if they get completely out of control, you only need to look at Venus. While carbon makes up less than 1 percent of Earth's atmosphere, on Venus it's about 96.5 percent.

Venus is closer to the sun than Earth is, which makes it hotter to begin with, but its atmosphere traps so much of the sun's heat that the average surface temperature is 864°F (462°C), compared to about 57°F (14°C) on Earth. (That's hot enough to melt lead.) Venus's greenhouse-like atmosphere is why, despite being farther away from the sun, it is so much hotter than Mercury, which is *only* about 332°F (167°C), but has a very faint atmosphere. Some scientists think Venus might have had water at one point, but whatever was once there evaporated because of the incredible heat.

Venus is an extreme example. Earth will never get quite that bad from human-made climate change. But that's because moving toward that kind of climate would kill off all of humankind long before we could produce enough emissions to get the amount of carbon in the atmosphere up to the Venus level. So that's the "good" news.

various factors, including oxygen produced by bacteria and other organisms, gradually cooled off the atmosphere. Dinosaurs lived for about 165 million years on a planet cold enough for giant reptiles to survive, but it was still a lot hotter than it is now. The Ice Age that arrived about 2.5 million years ago was a lot colder than the world we live in today, with glaciers expanding so far south that they covered nearly one-third of the planet. Animals like mammoths and giant ground sloths thrived.

The planet has gone through a lot of other changes over the past few million years. For example, most of Earth's land was once one supercontinent that later broke up into the seven continents and some of the thousands of islands we have now. Several mass extinctions have happened as well. Today's species represent less than one-tenth of 1 percent of all the organisms that have ever lived on the planet.

One of the important things to remember about the current climate crisis is that we're talking about how our climate compares to the rest of the time that human beings have been around. For the past few hundred thousand years—including the ten thousand or so that have seen some version of human civilization—Earth has been fairly cool, and the climate has stayed pretty stable. Sometimes it's been a little hotter or a little colder, but the changes haven't been extreme.

Until recently. And that's the problem.

What's happening to the climate now isn't part of any natural cycle; scientists have traced its origin to the beginning of the Industrial Revolution in the eighteenth century. Since then, manufacturing, agriculture, transportation, and other aspects of human civilization have changed drastically in a relatively short amount of time. And if we want to keep living the kinds of lives we do now, we're going to need to make some different choices.

How We Started Really Changing the Climate

Think for a moment about how any major American city is described in books or movies set in the late nineteenth or early twentieth centuries. A sky with clouds of black pollution and people coughing coal dust into handkerchiefs. Huge factories packed with workers. Trains and cars moving people around the city. The same pollution that was making the air so dirty was also starting to make the climate hotter.

For most of human civilization, we didn't have a major impact on the planet's average temperature. That changed with the increase in the greenhouse gases mentioned earlier. Carbon has been the main culprit, and humans have pumped a lot of it into the atmosphere by burning fossil fuels (oil, coal, and natural gas).

Earth would be a very different place without carbon. All life forms on the planet—from humans to fish to plants to bacteria—are carbon based. Humans and other animals breathe in oxygen and breathe out carbon dioxide, while plants absorb carbon dioxide and produce oxygen through a process called photosynthesis.

Until the Industrial Revolution, the amount of carbon that people released wasn't such a big deal. They released some by burning wood or lumps of coal to stay warm. Burning coal goes back to at least 1000 BCE and burning wood far longer than that. Things changed in the mid-1700s, however, starting in Great Britain. The invention and widespread adoption of the steam engine was one major factor. Steam engines operated by burning coal to heat water, and they could provide power more efficiently than manual labor could. This allowed people to develop industrial-scale machines.

Another factor was British inventor Abraham Darby's development of a new kind of coal-fired blast furnace to smelt iron, an important material for manufacturing and construction. As people invented new machines and expanded industries, those endeavors needed a lot more energy to run and a lot more fuel to produce it. In addition to coal, people turned to another fossil fuel—oil—to keep up with the demands of production.

So, what are fossil fuels, and why are they a problem? Millions of years ago, when ancient life forms such as plankton and microscopic animals died, they became fossils. They sank to the bottom of the sea, where they were buried under mud and other sediment. Millions and millions of years of heat and pressure turned the remains of these ancient organisms into deposits of what would one day become fuel. Coal formed from plant material, while crude oil and natural gas came from the remains of marine microorganisms. The carbon inside those fossilized organisms remained in the coal and oil they became. Burning these fuels for energy releases that carbon (and other greenhouse gases) in the form of emissions, and the atmosphere absorbs it. Because fossil fuels are so deep underground, the processes people use when drilling for oil, fracking for natural gas, and mining for coal require a lot of energy too.

Like a lot of issues related to climate change, the problem of carbon emissions is (pardon the pun) a matter of degrees. In 1750, when the Industrial Revolution was just firing up, people released fewer than 3 million tons (2.7 million t) of carbon emissions into the atmosphere per year. That might sound like a big amount (and it was for the time), but by the start of the twentieth century, humans released nearly 600 million tons (544 million t) of carbon emissions annually. In just a century and a half, the amount of carbon entering the atmosphere *per year* had gone up more than two hundred times. More than one-third of all the carbon currently in the atmosphere got there just in the past century.

The increase is due to the spread of industrialization around the globe. When industrialization was new, just a few areas of Great Britain were burning large amounts of coal, but technology soon spread to more countries and to more places within those countries. Steam engines that burned coal helped trains and boats carry people over long distances. By the 1850s, oil drilling had been established in the United States, and more and more people were using oil to light lamps. (This replaced whale oil, which was a big plus, as several whale species were at serious risk of being hunted to extinction.) The 1878 invention of the lightbulb caused a spike in demand for fossil fuels because the electricity for the bulbs came from burning oil. The introduction of the automobile a few years later increased demand even more.

Over time, industrial technology improved the lives of millions of people. However, these improvements have come with a cost: the rise of global temperatures. The challenge for the next few years will be to reduce that cost by adapting our standard of living to be safer and more environmentally friendly.

A BIG FRACKING PROBLEM

Natural gas has become more common recently, with production in the United States jumping 11 percent in 2018 from 2017 levels, which were already a record. Replacing other fossil fuels with natural gas, however, has both pluses and minuses.

On the plus side, natural gas only emits about half as much carbon as coal does, and the boom in gas development has allowed the United States to reduce the amount of coal it uses. Gas also burns more cleanly in cars than oil does, with fewer emissions. And because so much natural gas can be produced domestically, it has reduced (but by no means eliminated) the United States' dependence on importing oil from overseas, which has lots of potential geopolitical advantages.

On the other hand, the fracking process that's used to get the natural gas from underground shale is less than great. It involves drilling deep underground and then blasting a mix of water and chemicals in a high-pressure stream to break up the rock where the gas is stored. The released gas then flows into the well. This drilling and blasting are often done horizontally (drilling down and sideways, instead of just down), which can break up rock across a wide area.

In addition to the dangers of operating heavy machinery, workers at fracking sites are exposed to silica dust, hydrogen sulfide, and other toxic chemicals.

Fracking has other problems. It uses a lot of water, and it contaminates groundwater with fracking chemicals. The destabilizing of underground rock has been linked to earthquakes and sinkholes. In terms of climate, while the gas releases less carbon than oil or coal do when it burns, the fracking process releases methane that was stored underground into the atmosphere, which undoes some of the benefits. This is why natural gas is often talked about as a bridge fuel—a short-term option that still needs to be replaced.

People already understood that coal and oil emissions were a problem during the Industrial Revolution. They just weren't talking about it in terms of climate. Air pollution spread diseases and caused sometimes-deadly breathing issues. In the late 1800s, London was trying to solve its new "smoke problem," as bronchitis caused by air pollution had become the city's leading cause of death. Factories releasing pollutants into the water didn't help things either.

By the early twentieth century, cars were becoming much more common; airplanes followed suit within a few decades. In the closing years of the century, countries like China and India—each of which was already home to more than a billion people—started industrializing very quickly. By that point, scientists already understood the importance of reducing greenhouse gas emissions, but it was hard enough just to stop the amount from growing every year, let alone shrink it. By 2005, China had passed the United States as the planet's top carbon emitter, and by 2007 the developing world had passed the already-industrialized countries in terms of total annual emissions.

Unfortunately, that means the carbon emissions per year in the twenty-first century have far surpassed the figures from the early twentieth century. According to the Global Carbon Project, people set yet another less-than-ideal record in 2018, emitting an estimated 40 billion tons (36 billion t) of carbon. China, the United States, and India, in that order, were responsible for the largest amounts of carbon emissions that year. In terms of carbon emissions per person, however, the United States ranks higher than China and India.

And that's just carbon, the most common of the greenhouse gases. While industry and transportation generate a lot of carbon, animal farming does the same for methane. In the United States in 2017, carbon dioxide made up about 82 percent of greenhouse

WHERE THE CARBON COMES FROM

The United States is among the biggest carbon polluters in the world, but it's hardly the only one. The European Union, through its European Commission, released a comprehensive 2018 report on carbon emissions from fossil fuels by country. Here's what it found for 2017 (the most recent year for which it had full data) and how it compared to five years earlier. At least some of the top-ten countries, including the United States, managed to reduce their emissions in that span.

Most Total Carbon Emissions by Country 2017		
Country	Carbon emissions (kilotons)	Global share
China	10,877,217	29.3 percent
United States	5,107,393	13.8 percent
India	2,454,773	6.6 percent
Russia	1,764,865	4.8 percent
Japan	1,320,776	4.8 percent
Germany	796,528	2.1 percent
South Korea	673,323	1.8 percent
Iran	671,450	1.8 percent
Saudi Arabia	638,761	1.7 percent
Canada	617,300	1.7 percent

gases emitted and methane just 10 percent. But methane is between twenty and thirty times as powerful as carbon when it comes to trapping heat in the atmosphere. So, that relatively small percentage of methane could actually do more damage than the much larger percentage of carbon. According to NASA, methane gas emissions are up more than 150 percent since the start of the Industrial Revolution and, just like carbon emissions, they've gone up a lot in the past decade. Along with that, some of the same processes that dump carbon or methane into the atmosphere also emit nitrous

Most Total Carbon Emissions by Country 2012

Country	Carbon emissions (kilotons)	Global share
China	10,256,379	28.9 percent
United States	5,273,584	14.9 percent
India	1,988,738	5.6 percent
Russia	1,799,085	5.1 percent
Japan	1,289,286	3.6 percent
Germany	803,978	2.3 percent
South Korea	628,583	1.8 percent
Iran	586,811	1.7 percent
Canada	576,608	1.6 percent
Saudi Arabia	530,890	1.5 percent

Source: Marilena Muntean, Diego Guizzardi, Edwin Schaaf, Monica Crippa, Efisio Solazzo, Jos Olivier, and Elisabetta Vignati, "Fossil CO_2 Emissions of All World Countries," Publications Office of the European Union, November 22, 2018, https://publications.jrc.ec.europa.eu/repository/handle/JRC113738/.

oxide, water vapor, and fluorinated gases that all play a role in the greenhouse effect.

When discussing climate change, scientists often talk about a positive feedback loop, in which the effects of warming the planet create conditions that warm the planet even more. For example, as the global temperature goes up, more of the ice on Earth melts, which releases more of the methane trapped by the ice into the atmosphere. That makes the climate hotter, melting more ice, which makes the climate even hotter, and the loop continues. And as the oceans get warmer, more of the overheated water evaporates into water vapor, which also hangs out in the atmosphere and traps more heat. And so on.

Positive feedback loops aren't the only things putting stress on the planet's climate. As people cut down forests, there are fewer trees to take in carbon, which means more of the gas stays in the atmosphere. The burning of some of that wood increases emissions even more.

And as the number of people on Earth grows, there are more people using resources and producing greenhouse gases, which in turn creates more warming. (While every person adds to the emission issue, the amount of emissions per person varies by country; those in industrialized countries usually add more than those in the developing world.) Earth's population is also increasing dramatically. It took one hundred and twenty-three years for the world's population to grow from one billion to two billion, but only seventy-two years to go from two billion to six billion. That means dramatic emissions cuts are needed to improve the situation. All of these factors mean that the longer we go without making serious efforts to address the climate crisis, the less time we will have to fix things.

Some people talk about the climate crisis in terms of saving Earth, but the truth is the planet itself will survive. It was around for billions of years before we came along, and it's going to be around long after us. A better way to talk about addressing climate change is that it means trying to save humanity and the other life on Earth—which starts with reducing greenhouse gas emissions.

How Do Scientists Know All This?

The fact that greenhouse gases are heating the planet isn't some new discovery. Serious discussion of the science behind climate change goes back at least to the nineteenth century.

In 1859, Irish physicist John Tyndall conducted a series of experiments to test which "colorless and invisible gases and vapors" do the best job of absorbing heat. Not only did he identify carbon dioxide, water vapor, and ozone as gases that absorb heat better than the overall atmosphere does, but he figured out that other atmospheric gases like oxygen, nitrogen, and hydrogen are worse than average at absorbing heat. Tyndall's 1859 identification of those gases was a big scientific breakthrough, and he's often credited as the first to prove the greenhouse effect (though that term wasn't used quite yet).

But Tyndall wasn't the first to study it. French scientist Joseph Fourier proposed in the 1820s that Earth's size and distance from the sun meant that the sun alone couldn't explain how warm the planet was. He theorized that the atmosphere might trap heat and insulate the planet, and he was right. In 1856, Eunice Foote linked carbon to that idea, demonstrating that a cylinder of air with carbon dioxide pumped into it stayed hotter longer after exposure to sunlight than a cylinder without the extra carbon. Unfortunately, Foote didn't personally present her findings at the conference of the American Association for the Advancement of Science (one of her male colleagues did it instead, though it's unclear who made that decision), and the gender bias of the time might explain why her work didn't become well known until much later. But she was right too.

Swedish chemist and physicist Svante Arrhenius, winner of the 1903 Nobel Prize, published a paper in 1896 titled "On the Influence of Carbonic Acid in the Air Upon the Temperature of the Ground." In it, Arrhenius discussed how certain gases in the atmosphere absorb heat. Many scientists consider the paper to be the first study to quantify the greenhouse effect. Arrhenius was trying to figure out what caused the Ice Age and made the connection between the amount of key gases in the atmosphere and changes in Earth's temperature. Just as he calculated that lower levels of carbon would make the planet colder, he theorized that higher levels would make things warmer. He actually saw increasing carbon levels as positive and maybe even something worth trying to do on purpose, but he also lived in a colder time. And in Sweden, which was pretty cold to begin with.

These scientists' work on climate change focused on the big picture. The study of short-term, year-by-year emissions began in the second half of the twentieth century.

One important step in tracking climate change was to consistently measure the amount of carbon in the atmosphere. In the late 1950s, scientist David Keeling and the California-based Scripps Institution of Oceanography set up a climate-monitoring station in Hawaii, on the slope of the Mauna Loa volcano. They chose the volcano because it was more than 11,000 feet (3,350 m) above sea level and because the surrounding lava kept the station away from vegetation, soil, and human activity—any of which might affect the accuracy of the measurements.

Starting in 1958, the observatory measured the concentration of carbon dioxide in the atmosphere—and it took only a few months to determine that the numbers were slowly going up. The US government got involved a few years later when the National Science Foundation started funding the research. In 1963, the foundation issued a warning about the dangers of the increased carbon in the atmosphere. When computer technology started advancing in the 1960s, researchers were able to build computer models that used the collected data from the past and present to start predicting the future of climate change.

The Mauna Loa observations were groundbreaking, and the observatory's ongoing research is important, but it's hardly the only source of data. With the risk becoming clear, more scientists around the world started studying the amount of greenhouse gases in the atmosphere and their impact on the global climate. Dozens of stations began conducting similar research. They used satellites, radiosonde transmitters attached to weather balloons, radar technology, and other tools to collect more data. Their findings have backed up the steady increase American scientists were seeing, and climate science became an increasingly important field of study.

The data from Mauna Loa and other stations showed that the amount of carbon in the atmosphere was going up consistently and rising more dramatically as time went along. That ever-growing curve from 1958 to the present is nicknamed the "Keeling Curve," after the founder of the observatory. As we learned more about the amount of carbon that was in the atmosphere before 1958, the curve started to look even more dramatic.

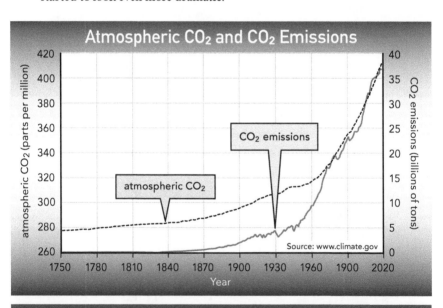

The Keeling Curve demonstrates a sharp rise in atmospheric carbon dioxide that correlates strongly with the amount of carbon dioxide emitted by human activities. Until the Industrial Revolution, atmospheric carbon dioxide levels were fairly stable.

HOCKEY ANYONE?

In a famous scene from the Oscar-winning 2006 documentary *An Inconvenient Truth*, former US vice president Al Gore introduced the audience to a chart tracking the amount of greenhouse gases over time. To emphasize the scope of the problem, he projected the chart onto a wall and used a scissor lift (a kind of mechanical ladder) to show the crowd how much the chart goes up as it moves into the present day and into the future.

This graph, which extends the Keeling Curve far back into the past, is often called the "hockey stick" because of the shape formed by a really long stretch of minimal change shooting up at the end. It was introduced by climate scientists Michael Mann, Raymond Bradley, and Malcolm Hughes in 1999. Because the graph was dramatic and scary, and because similar graphs featured prominently in some critical IPCC reports, climate deniers worked overtime to try disproving it. They targeted the scientists involved as part of the "Climategate" email hack (see page 42), and climate deniers in Congress launched investigations into the scientists' research, demanding notes and testimony. Part of the "controversy" focused on the older numbers, which weren't as well defined as the modern ones, and where there was some fluctuation between different studies (none of which disputed the fact that carbon emissions had been increasing since

To figure out what the global climate was like before humans began measuring it was a trickier matter. Luckily, long before anyone started making such measurements, Earth was already keeping records. These natural records aren't as precise as the ones captured by technology, but they still do a good job of tracking big-picture patterns over time.

Ice cores are one important resource scientists have studied to figure out long-term climate patterns. The amount of ice that forms every winter at the north and south poles partly depends on how hot or cold the atmosphere is, which impacts how much water freezes. The yearly pattern of seasons becomes visible in layers of ice stacked on top of one another, with thinner layers produced during warmer periods and thicker ones in colder periods.

industrialization). Deniers figured if they could undermine part of the data, they could convince people the whole chart was questionable and that human-made climate change was a hoax.

Of course, they were wrong. Also, as scientists tested the chart's numbers, and as they found more old data to study, they produced more "hockey stick"-style graphs with more detailed data—and with longer segments going back through time. Deniers had taken their best shot on goal, and scientific research and data made a crucial save.

Former US vice president Al Gore speaks at a recent climate summit. Since he left office in 2001, Gore has continued to speak about and warn of the imminent dangers posed by climate change.

Because the north and south poles are isolated, the long-term data stored in the ice there have been uncontaminated by human activity for millions of years. By drilling deep and removing long vertical tubes of ice, scientists can study clues trapped in the ice over millennia. Bubbles frozen inside the cores contain samples of the air at the time, and comparing bubbles from different layers can show changes in the makeup of the atmosphere. So can studying things like ash, dust, and other particles frozen in those cores. In Antarctica in 2017, scientists collected an ice core containing 2.7 million years' worth of ice. Among other things, it showed that the amount of carbon in the atmosphere never passed three hundred parts per million until recent post-industrial climate-change. For context, by 2020, the amount of carbon in the atmosphere passed four hundred parts per million.

Scientists also collect samples from the seabed all over the world. While annual weather patterns don't impact the seabed as directly as they do the ice, changes in the layers of sediment can provide clues about which elements were common in a specific period. For example, based on which plant and animal fossils a certain layer holds, scientists can determine what the water temperature was like at that time. Also, comparing seabed layers from different places from the same period provides evidence about ocean currents (which are affected by the climate), and the layers can hold some of the same kinds of particles found in ice cores. Tree rings are also indicators of climate over time, though they're less reliable than polar ice and seabeds because they're more easily impacted by local weather events, such as fires or droughts, that weren't necessarily global.

In the 1980s, temperatures had gone up noticeably, so climate change and the science behind it started to get a lot more attention. In 1983, the US Environmental Protection Agency (EPA) called global warming "a threat whose effects will be felt within a few years." When 1988 became the hottest year in recorded history—remember, that record has been broken many times since then—the public started to understand that the increased heat might have contributed to that year's serious fires and droughts. That June, NASA scientist James Hansen testified before Congress that the science was "99 percent" clear that global warming was happening and that humans were causing it. A cautious scientist by nature (just like most climate scientists), Hansen said he couldn't prove which specific fires and droughts were directly caused by climate change, but he noted that the number of them was telling and that they were exactly the kind of events that would become more common if global warming continued to get worse. He also presented three possible future warming scenarios based on the research to that point. Each predicted how hot Earth would get if different quantities of greenhouse gases were released in the years ahead, with the outcomes for the planet better or worse depending on those amounts.

The science Hansen and others presented was troubling enough that countries finally started taking action. Then presidential candidate George H. W. Bush (who won the election in November) promised to use the "White House effect" to fight the greenhouse effect. How important was Hansen's testimony? A fellow atmospheric scientist, Michael Oppenheimer, said at the time that it "almost overnight" made climate science a national political issue. It was a big deal.

A TAXING SITUATION

In November 2018, voters in Washington state rejected Initiative 1631, which would have been the first state-level carbon tax in the United States instituted directly by voters. This happened despite Washington being considered a fairly green state whose governor, Jay Inslee, briefly mounted a 2020 presidential campaign with addressing climate change as his main issue.

If Initiative 1631 had passed, the state would have charged fees to polluters who produce greenhouse emissions. The money from those fees would have been spent on conservation efforts. Opponents spent more than $30 million to defeat the measure— the most ever spent on a ballot initiative in the state. Most of the money came from just a few fossil-fuel companies, including usual suspects Chevron, BP, Shell, and Koch Industries, as well as petroleum-industry groups like the Western States Petroleum Association.

To convince voters to reject the measure, these businesses made clear that they intended to pass the cost of the fees directly on to consumers, raising the price of gasoline and home heating gas (though by only about ten and fifteen cents per gallon, respectively). That tactic worked. Ultimately, 56 percent of those who showed up to vote said no to the measure.

In 1992, the United States was one of the nations that negotiated the United Nations Framework Convention on Climate Change treaty, with the goal of limiting the amount of greenhouse gas in the atmosphere. The treaty wasn't binding, so countries couldn't be forced to hit their targets, and the fact that the climate has gotten a lot hotter since then proves it wasn't as effective as it could have been. Still, it was an important first step for countries to at least agree that there was a problem, think about how to address it, and promise to try.

Just as importantly, in 1988, the United Nations Environment Programme and the World Meteorological Organization started the IPCC. The IPCC brings together experts from around the world—195 countries are members—to share the best objective research on climate science. It also issues reports about the future impacts of climate change, including the kinds of effects we'll talk about in the coming chapters. The IPCC doesn't conduct much research itself, but it reviews, compiles, and publishes research by thousands of top international scientists, and it has made some important predictions.

Speaking of thousands of scientists, one fact that gets thrown around a lot when talking about climate is that "97 percent of scientists" agree that climate change is happening and that humans are causing it. That number comes from a series of studies that looked at the peer-reviewed papers scientists published about climate change. Serious scientific institutions and publications use peer review, a process in which other scientists who are experts in the specific field check the work before it's published to make sure it's not plagiarized, falsified, inaccurate, sloppy, or otherwise not up to snuff.

While they sometimes have different predictions for exactly how much the temperature is likely to rise and for how much time we have left if we're going to avoid the absolute worst outcomes, scientists agree on the cause and seriousness of the problem. Various consensus studies, which are essentially peer-reviewed studies of peer-reviewed studies, looked at the research on climate change and found a scientific consensus, or general agreement, of at least 97 percent of scientists. Considering that it's hard to get 97 percent of people to agree on much of anything, that's a pretty significant number. And that's just the percentage of studies that argued human activity is driving climate change. The percentage of scientists who agree that climate change exists is even higher, with the consensus only increasing among researchers with more experience and expertise.

The last paragraph of a 2016 study about the consensus, published in the journal *Environmental Research Letters*, put it like this: "From a broader perspective, it doesn't matter if the consensus number is 90 [percent] or 100 [percent]. The level of scientific agreement on AGW [anthropogenic, or human-made, global warming] is overwhelmingly high because the supporting evidence is overwhelmingly strong."

As we've discussed in this section, climate science goes back quite a long time, and scientists are sure about what's happening and why. There is no scientific controversy about it. Unfortunately, the same can't be said about political controversy.

The Science Is Clear, But . . .

Just a generation ago, the political debate about climate change was mainly an argument about the best approach. Should it involve governments setting rules and regulations to improve the situation, or should private businesses come up with market-based solutions? In the United States specifically (most other countries got over this long ago), however, the national debate has morphed into a fight over whether to address the problem at all, with facts on one side and conspiracy theories and science denial on the other.

The fossil fuel industry had a lot to lose if governments started acting on the science, since that could limit the use of its products and cost it money. It has spent decades trying to convince people that climate change was anything from part of a natural cycle to a hoax cooked up by scientists or journalists or China or the bogeyman.

Oil and coal companies were hardly the first to use such tactics. Tobacco companies publicly denied for years—even when executives were under oath in court—that cigarettes and other tobacco products caused cancer. As part of a 1998 lawsuit, companies had to release old

INCENTIVIZED STUPIDITY

One of the reasons climate denial has become so widespread is that many American politicians, mostly Republicans, have spent years turning it into a partisan issue.

There are plenty of examples, but perhaps nobody in government loves denying science more than GOP senator James Inhofe of Oklahoma. In 2012, he published a book that promoted anti-science conspiracies. Then during an infamous 2015 speech, he held up a snowball to prove climate change wasn't real. Because it snowed. In February. When it often snows in Washington, DC.

Unfortunately, Inhofe's climate denial didn't stand in the way of his serving as chairman of the US Senate's Environment and Public Works Committee in 2003–2006 and again in 2015–2017. In that role, he frequently stopped legislation that could have helped get climate change under control, such as stronger emission standards.

There are a few possible explanations for Inhofe's attitude and actions. First, it could be he just isn't smart enough to know the difference between weather and climate, or he doesn't care enough to figure it out. Or it might have something to do with the fact that the oil and gas industry has consistently been Inhofe's biggest source of funding, donating hundreds of thousands of dollars to his campaigns. Or both. Neither of these options is encouraging when the guy has such a big say in environmental policy.

documents that showed they had known the cancer danger was real since at least 1959, but intentionally hid it from the public and lied about it. The US government sued the big tobacco companies in 1999 and won in 2006, with the companies forced to run ads admitting the danger. By that point, nearly half a million Americans per year were still dying from smoking-related illnesses, even though the percentage of Americans who smoked was less than half of what it was fifty years earlier.

Companies that produce fossil fuels and other industries that use or invest in those fuels took similar steps. As far back as the 1970s and 1980s, Exxon—one of America's largest oil companies (which has since become ExxonMobil)—had its own scientists studying climate change, with results that matched everyone else's. But rather than act on that information, internal documents showed that in 1989 the company shifted to a corporate strategy to emphasize the "uncertainty" and promote the lie that climate change was just a possibility, rather than a guarantee.

Exxon was also one of several energy companies that created and funded lobbyists and phony "junk science" groups to tell the public that climate change wasn't a serious threat, wasn't caused by people, or wasn't even real. Exxon spent more than $31 million on those efforts between 1998 and 2005. In 2018, oil company Chevron's attorneys admitted in court what the company had denied for years— that there's no debate about the scientific reality of climate change. But it took the cities of San Francisco and Oakland suing five big oil companies (ExxonMobil, Chevron, BP, Shell, and ConocoPhillips) and making their executives testify under oath to get that admission. And those are just a few examples.

Oil and coal companies joined manufacturing companies to set up lobbying groups and think tanks with innocent-sounding names like the Global Climate Coalition, the George C. Marshall Institute, and the Heartland Institute, which released reports and articles

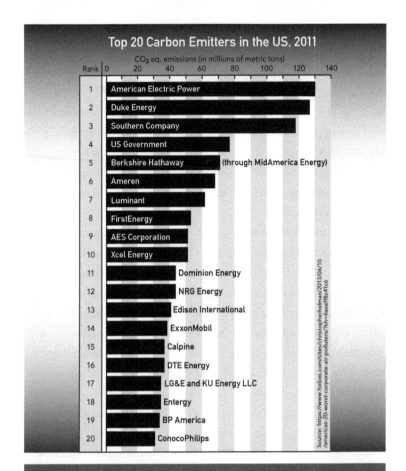

Top 20 Carbon Emitters in the US, 2011

CO$_2$ eq. emissions (in millions of metric tons)

Rank	Company
1	American Electric Power
2	Duke Energy
3	Southern Company
4	US Government
5	Berkshire Hathaway (through MidAmerica Energy)
6	Ameren
7	Luminant
8	FirstEnergy
9	AES Corporation
10	Xcel Energy
11	Dominion Energy
12	NRG Energy
13	Edison International
14	ExxonMobil
15	Calpine
16	DTE Energy
17	LG&E and KU Energy LLC
18	Entergy
19	BP America
20	ConocoPhilips

Source: https://www.forbes.com/sites/christophehelman/2013/06/10/americas-20-worst-corporate-air-polluters/?sh=6aeaff8c41c6

This graph shows the top twenty carbon emitters in the United States in 2011. Since then, while the companies may have changed rankings, it remains true that private fossil fuel companies and the government, particularly the US military, are responsible for the vast majority of US carbon emissions.

promoting anti-science nonsense. They also created groups to protest climate action, spread conspiracy theories, and organize campaigns to try to discredit serious science. By setting up these anti-science organizations, the fossil fuel industry successfully spread doubt among enough of the public to create the perception of controversy, all for the sake of maintaining its profits.

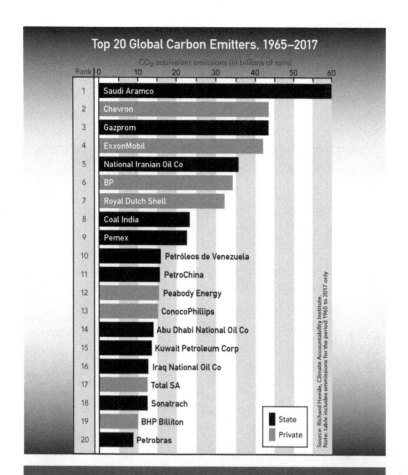

Top 20 Global Carbon Emitters, 1965–2017

CO₂ equivalent emissions (in billions of tons)

Rank	Company	
1	Saudi Aramco	State
2	Chevron	Private
3	Gazprom	State
4	ExxonMobil	Private
5	National Iranian Oil Co	State
6	BP	Private
7	Royal Dutch Shell	Private
8	Coal India	State
9	Pemex	State
10	Petróleos de Venezuela	State
11	PetroChina	State
12	Peabody Energy	Private
13	ConocoPhillips	Private
14	Abu Dhabi National Oil Co	State
15	Kuwait Petroleum Corp	State
16	Iraq National Oil Co	State
17	Total SA	Private
18	Sonatrach	State
19	BHP Billiton	Private
20	Petrobras	State

State
Private

Source: Richard Heede, Climate Accountability Institute.
Note: table includes emmissions for the period 1965 to 2017 only

This graph shows the top twenty carbon emitters globally. Government-owned and private fossil fuel companies produce the vast majority of carbon dioxide and other greenhouse gas emissions.

They also spent a lot of money on the campaigns of anti-science politicians who would work against "green" (environmentally friendly) initiatives and tell voters that climate science was a hoax. (See the sidebar on page 42.) Government officials and "experts" funded by the industry have spread lies and half-truths in every election cycle for three decades. These messages were supported by partisan outlets such as Fox News and talk radio, effectively delaying serious action while the problem only got worse.

One cause for optimism is that, despite all these efforts by deniers, most Americans haven't been fooled. A November 2018 poll by

A HACK OF A DIRTY TRICK

One way in which liars try to convince people climate science isn't real is to claim that scientists are somehow all involved in a mass conspiracy for . . . some reason. Some dishonest actors have gone to great lengths to try proving that this fake conspiracy exists.

Before the Copenhagen Summit on climate change kicked off in 2009, hackers stole thousands of emails and documents from the Climate Research Unit at the University of East Anglia in England. Soon, blogs and other sites popular with science deniers began posting a handful of those emails—without including the surrounding conversations and intentionally picking messages that could easily be taken out of context—and tried to create a phony scandal they called "Climategate."

In one case, a scientist was quoted as writing, "We can't account for the lack of warming at the moment and it is a travesty that we can't." If that was all you read, it might have sounded like he was admitting there was a key gap in the research supporting climate change. In reality, he was talking about the need for a better way to measure specific short-term changes. Other emails discussed why records showed some cooling in the 1960s and how to change the data . . . because it was known by then that tree rings (which were used for data at that time) were not as reliable on their own as people once thought. No conspiracy, just scientists discussing how to do their jobs better and improve the data.

Proving that "Climategate" was pure fiction wasn't difficult, since the emails were so clearly being misrepresented and made more sense once they were shown in context. But the hackers and deniers who tried to make it a thing managed to get the "scandal" into the news briefly. They also got governments to launch several investigations, none of which found any wrongdoing by scientists or cast any doubt on the scientific consensus about climate change. However, there's little doubt that some people who heard about the "scandal" and the investigations—but didn't pay attention to the details or learn how it all turned out—fell for their scam.

Monmouth University found that four in five Americans agree that climate change is real (that it wasn't five in five is still a problem, of course). While there is a clear partisan divide about the issue, with about 92 percent of Democrats and 78 percent of independents agreeing that climate change is real, a majority of Republicans—64 percent, so nearly two out of three—also got it. Most Americans in that poll—and in several others from 2018 and 2019—said they want the government to do more about climate change. In a 2019 study by the Pew Research Center, 59 percent of Americans said that climate change was already having a great impact on their own communities—seeing is believing, after all. A September 2019 CBS poll found two in three Americans consider climate change a serious problem or a crisis.

More good news is that those numbers had increased since 2015, meaning people were starting to understand the importance of the issue. Hopefully those numbers will keep going up. Unfortunately, the efforts of climate deniers over the last decade, particularly in the United States, cost a lot of time while people were fighting about whether the problem existed instead of focusing on how best to solve it.

This can't be stressed enough. Anyone trying to convince you that climate change isn't happening or that it isn't caused by human activity falls into one of two categories. At best, they might have fallen for misinformation that someone with bad motives fed them. At worst, they know better and are blatantly lying. It's important to remember that most climate deniers are intentionally dishonest, whether it's to line their own pockets or just to be jerks or trolls.

Now that we've covered the basics of climate change and how it works, we can start talking about what it means for the planet, both now and in the future.

CHAPTER 2
A CHANGE IN THE WEATHER

What changed in the US with Hurricane
Katrina was a feeling that we have entered a
period of consequences.

—former US vice president Al Gore in 2006, a year before winning
the Nobel Peace Prize for his work on climate change

SOME UNUSUAL WILDLIFE videos and photos from Australia
went viral on social media in August 2019. A lot of people there (and
around the world) had never seen kangaroos hopping around in a
snowstorm before. Other viral clips showed koalas in trees covered
in the white powder and wombats leaving tracks as they trudged
through it.

While some of the more mountainous parts of the country get
snow pretty often, most of Australia is much too hot for that—even
in winter, the country averages temperatures in the high fifties (or in

the teens in Celsius), and large parts of it are even warmer. During the 2019 snowstorms, some towns just a few hours' drive from Sydney were covered in snow for the first time in decades. People old enough to have graduated from college had never seen the stuff there in person. Along with that, the storms brought winds of up to 70 miles (112.7 km) per hour. The winds canceled hundreds of flights, damaged homes, broke off a huge chunk of a pier, and caused numerous other problems.

Snowstorms in Australia have been rare enough in the past century and a half that the weather was a huge news story there. Meanwhile, on the other side of the world, a heat wave was making headlines that same month.

Phoenix, Arizona, one of the largest and fastest-growing cities in the United States, is situated near the edge of the Sonoran Desert. The city is usually hot in August (its basketball team is named the Suns for a reason). But 2019 was even hotter than normal, breaking many heat records for the city. According to the National Weather Service, the 112°F (44.4°C) temperature on August 20 set an all-time high for that date. So did the 114°F (45.6°C) temperature on August 21. And the record-breaking heat just kept coming. Of the ninety-two days from June to August, forty-three were the hottest they'd been since 2000. Unfortunately, that same record heat also led to a record number of heat-related deaths for the third year in a row.

Those two situations on opposite sides of the planet are just a couple examples of the kind of extreme conditions that climate change is already bringing. Every year, heat waves, record rainfall, hurricanes, and other intense weather keep increasing around the world.

Not every example of extreme weather is directly linked to climate change, but climate change makes those events much more likely and much more intense. To borrow an example from chapter 1, we've known for decades that there's a strong link between smoking and

REMEMBERING A HEAT WAVE

Chicago went through an awful heat wave during the summer of 1995. The temperature got up to 106°F (41°C), with temperatures remaining above 80°F (26.7°C) at night. The humidity made it feel even hotter—the kind of air that felt like you were wearing a heavy wet jacket—and it took just a few seconds of standing outside to start sweating. With humidity trapped low in the atmosphere, there was very little wind, which otherwise could have provided a breeze to cool things off a bit. Instead, the heat index hit as high as 125°F (51.7°C), which was incredibly unpleasant. (I can personally say that it was a rough time to work a summer job at an outdoor concert venue. It required drinking huge amounts of water all night just to avoid overheating.)

People lucky enough to live in homes with air conditioning understandably used it, putting pressure on the region's electrical grid. Not everyone was so lucky. Hospitals and ambulances were overburdened trying to treat people dealing with heatstroke, dehydration, shock, and other health problems.

In neighborhoods where buildings didn't have air conditioning or reliable water and where it was too dangerous to sleep outside, the heat turned deadly. In just five days, more than seven hundred Chicagoans died from heat-related causes. The city has since opened cooling centers and updated its electrical grid, among other actions, so that future heat waves don't become so deadly.

lung cancer. There are cases of people who've never even picked up a cigarette getting lung cancer, but smoking makes it a lot more likely and is behind a large number of cases.

The number of extreme weather events we've seen over the last several years would be impossible *without* the conditions caused by climate change. These events are some of the most visible impacts of what's already happening to Earth's atmosphere, and they'll keep happening—and probably get worse—if we continue along our current path. What seems extreme now could soon feel normal, which is a troubling proposition.

A Longer, Even Hotter Summer

Among all the types of weather that climate change is making more extreme, heat waves might be the most obvious example. As we covered in the last chapter, the planet as a whole is getting hotter on average. This means not only that hot days occur more often, but the periods of super-hot weather last longer.

In the United States, the number of heat waves per year tripled from an average of two in the 1960s to six in the 2010s. The federal government's US Global Change Research Program found that forty-six major US cities already have heat waves more often than they did fifty years ago, and the heat wave season—the period during the summer when they can happen—has become forty-seven days (more than a month and a half) longer in that same span.

The hotter climate causing hotter weather is one factor causing longer heat waves. Another is the impact the heat is having on the planet's jet streams. These are high winds in the atmosphere that help move weather systems around the planet. Lately, scientists have found that changes in the atmosphere are blocking waves in the jet stream. They're slowing down or even getting stuck in place when they should be moving—a traffic jam is one way of describing it. That means less wind movement, so hot air can hang around longer in one place— which can stretch a hot day or two into an extended heat wave.

On top of the heat, there are changes in humidity. Humidity is caused by evaporating water that stays in the atmosphere, making the air feel sticky and heavy. When the human body gets too hot, it cools itself by sweating—but that only works if the sweat can evaporate. On humid days, there's already a large amount of water in the air, so sweat evaporates more slowly, and the body isn't able to cool itself as easily. That unevaporated sweat makes it feel even hotter.

As the planet gets warmer, the atmosphere is able to hold more moisture, so humidity sticks around. That can be really dangerous. You're more likely to get heat exhaustion, or even heatstroke, on

YOUR HOMETOWN HEAT WAVE

While there is a range of predictions for how bad heat waves are going to get, you might be concerned mostly with how they'll affect the place where you live.

For those living in the contiguous United States (so not Alaska, Hawaii, or the territories such as Puerto Rico, Guam, or American Samoa), the Union of Concerned Scientists has created an online killer heat interactive tool. It lets you search by your city or county and the level of heat you want to know about (above 90°F [32°C], above 100°F [38°C], above 105°F [41°C], or "off the charts"). Based on the organization's research, which uses historical trends to model future warming, the site will show you an average from 1971–2000, as well as two predictions for the future (where we're currently headed and what mitigation efforts could produce). You can check it out at https://ucsusa.org/resources/killer-heat-interactive-tool.

days when the heat index is above 100°F (38°C). The heat index is a measurement that uses temperature and humidity to determine how the weather feels to the human body, beyond just what the thermometer says. Extreme heat and humidity can affect the functioning of organs like the heart and lungs, which can be deadly.

Pick a part of the world and people living there have probably had to deal with a record heat wave at some point in the last few years. Europe went through a terrible heat wave in the summer of 2003 that killed more than thirty-five thousand people. That wave's temperatures were topped in 2019, when two summer heat waves shattered records in Germany, Belgium, the Netherlands, France, and elsewhere. In 2018, at least sixty-five people in Japan died in just one week from heatstroke or other heat-related issues and more than twenty-two thousand were hospitalized.

In 2017, Australia had a record-hot summer, with parts of the state of New South Wales (which includes Sydney) hitting 113°F (45°C). It had its warmest December ever in 2018 and broke that record again in 2019. India and Pakistan endured a terrible heat wave that reached 123°F (50.5 °C) in June 2019, killing more than one thousand people. From Mexico to Canada, much of North America experienced a deadly heat wave in 2018. In Montreal, where the residents aren't used to extreme heat, temperatures of "only" 98°F (36.7°C) resulted in the deaths of sixty-six people. Russia lost more than fifty-six thousand people to a heat wave in 2010. The Middle East is particularly susceptible to extreme heat. One day in 2015, the Iranian city of Bandar Mahshahr hit 115°F (46°C), but humidity made the heat index an incredible 165°F (73.9°C).

How bad will heat waves get in the future? Well, that depends on how much more the planet warms. The US government's 2018 National Climate Assessment predicted that, across forty-nine of the country's largest cities (which include Wichita, Raleigh, and Omaha), extreme heat waves will kill at least nine thousand more people per year by the end of the century than they do now. Even that number assumes emissions stay generally where they are now. If they go up, so would those death tolls.

As bad as that is, it's also a fairly conservative prediction. The non-profit Natural Resources Defense Council released a 2017 study that predicted an average of nearly thirty thousand extra deaths per year in major US cities by the 2090s—a number almost twice as high as the total number of annual homicides in the United States and ten times as high as the number of people killed in the terrorist attacks of September 11, 2001. A 2019 study by the Union of Concerned Scientists is even scarier, predicting that parts of the country (mostly along the coast of the Gulf of Mexico) could experience 120 days per year—about four months—with a heat index above 100°F (38°C).

The Rain Must Fall

Along with increases in humidity, the extra moisture in the air causes more precipitation. The 2018–2019 rainy season was an unusually wet one in the United States, with three months in a row breaking records as the wettest in the country's history. Nearly 40 inches (101.6 cm) of precipitation (rain, snow, hail) fell in the United States during that time, almost 8 inches (20.3 cm) above average. Since 1900, the country's annual rainfall has gone up about 5 percent. That extra rainfall harms wildlife, causes floods, and increases erosion, among other effects.

The difference between climate and weather is crucial here too. The US climate overall is wetter. The numbers don't lie. As far as weather goes, though, not all parts of the country are getting more rain or snow.

If you live in New England, the odds are you're dealing with wetter weather than older generations there did when they were your age. If you're in the Southwest, on the other hand, your town is probably getting less rain than it used to get just a few decades ago. If you're in Montana, you've probably had to dig out of worse-than-normal snowstorms—in 2019, the state got more than 4 feet (1.2 m) of snow during the first week of October, an early storm that prompted the governor to declare a state of emergency.

That all matches up with the trend scientists are predicting worldwide. Wet areas are getting wetter, and dry areas are getting drier.

While scientists have used climate models based on existing data to predict these kinds of changes for a long time, we're past the point where these are merely predictions. Annual measurements are also verifying the numbers. A 2018 study, for example, used data measured by fifty thousand weather stations to find that parts of the eastern United States broke records for monthly rainfall 25 percent of the time between 1980 and 2013.

Climate change affects the amount of water in the air, leading to extra rain in some places and making water harder to find in others. Water covers about 71 percent of Earth's surface, and the five oceans—Pacific, Atlantic, Indian, Southern, and Arctic—account for more than 96 percent of that water. When the oceans get hot, some of that water evaporates. Think of how heating a tea kettle or boiling a pot of water on the stove causes some of the water to evaporate and how turning up the heat makes that process happen faster. With so much water on the planet, even small changes in global temperature can produce a lot more evaporation.

Because water vapor is itself a greenhouse gas, more of it in the atmosphere plays a role in the planet getting hotter, which in turn causes more evaporation and more water vapor (another good example of a feedback loop). For instance, an unusually warm winter in 2006 prevented Lake Erie from freezing over for the first time in recorded history. That made snowstorms in the area worse because the unfrozen water from the lake evaporated and came back as snow.

Precipitation is yet another example of climate change taking a normal process to the extreme. The water cycle—with liquid water evaporating, condensing in the atmosphere in cloud form, and coming back to Earth as rain (or snow, sleet, or hail)—is normal. What's changed is the amount of water that's evaporating and the amount getting dumped back on us.

Changes to winds are another concern. When jet-stream wind patterns get stuck in place and cause heat waves, they also cause weather systems to move more slowly and to stay in one area for a while. That means storms last longer in certain places because patterns that would normally dump rain over a wide area instead pour most or all of it across a smaller range.

Along with that, because of how weather patterns work, we're seeing precipitation moving away from the equator and toward the north and south poles. Areas farther north and south can expect to get more dramatic amounts of rain and snow, and areas near the equator and in the tropics can expect to get much drier. From 1900 to now, northern Europe and northern Asia have also gotten wetter, while the Mediterranean and northern Africa have seen less rain.

The same warming forces that are causing more heat waves are also leading to fewer cold waves as winters get shorter and generally milder. So, in places where the amount of snowfall is increasing, it's actually happening over fewer days. More snow over shorter periods affects how quickly snowpack—the hard snow that stays on the ground and builds up—forms and how soon it melts. This has an impact on everything from the length of tourist season in skiing areas to the survival of the numerous plants that rely on water from melting snow to grow.

While scientists can't say for sure that any one specific rainstorm or snowy day is the direct result of climate change, the overall increase in both the number of these weather events and the amount of precipitation they bring at once couldn't happen without it. As the planet gets hotter, there will be more wet days and more precipitation during those days. A team of US-led climate scientists put it bluntly in a 2019 study: the weather we're seeing would have a one-in-a-million chance of happening without climate change.

Water, Water Everywhere

The same month that Phoenix saw the record heat mentioned earlier, India was dealing with a brutal monsoon season that killed thousands of people. Every year, from late spring into early fall, the monsoon, a seasonal shift in the prevailing winds, hits parts of Asia. Though some people might think of a monsoon as a rainy season— and a rainy season does come along with the monsoon every year— the name really refers to that shifting wind. While other parts of the world have monsoons too and there is also a dry monsoon season, the most dramatic version is the wet monsoon season in Asia, during which humid air from the northern part of the Indian Ocean flows over India, Pakistan, Bangladesh, Thailand, Cambodia, Vietnam, and other parts of southern Asia.

The difference between the temperature of the ocean and that of the land causes the wind to shift. That shift brings huge amounts of rain with it, dumping about three-quarters of India's annual rainfall in just a few months. That rain is important for plants, people, and other animals that have adapted to it—and the flooding that accompanies it —every year. But here, too, climate change is messing with things.

Two of the three worst floods in western India happened during the 2018 and 2019 monsoons, and the amount of rain the monsoons bring has tripled over time. The 2019 storm was so bad that hundreds of thousands of people living in three states along the west coast of India—where they were used to monsoon-related storms and were prepared for heavy rain—needed to evacuate their homes and stay in relief camps until it was safe to return. They were relatively lucky. Thousands of other people across the region died in the floods, and more than two and a half million people were injured or permanently lost their homes.

Because about one-third of the world's population lives in areas affected by these monsoons, the storms have a big impact. With ocean temperatures changing when and how dramatically monsoon rains hit, the warming of the planet will cause major problems on this front.

Massive floods have also occurred in plenty of other places throughout the past decade. In 2013, Boulder, Colorado, got 14 to 17 inches (35.6 to 43.2 cm) of rain in a week—about three-quarters of the city's annual average—which caused the worst flood in that city's history. Much of central Europe also experienced extreme flooding that summer because of heavy rainfall. Tens of thousands of people in Germany, Hungary, Slovakia, the Czech Republic, Austria, and other countries needed to evacuate their homes in some of the worst flooding that region had seen in centuries. The Danube, Elbe, and other rivers flooded, covering the streets of entire cities.

When rivers or similar bodies of water take on too much rain at once, they can overflow their banks, even in places where dams or levees have been built to keep water under control. If there's too much

A TROUBLESOME LITTLE BOY

One obstacle to measuring exactly how much impact climate change has on precipitation is a weather pattern called El Niño—Spanish for "little boy." (The term was first used by South American fishers in the 1600s.)

El Niño happens every few years. When the trade winds over the eastern Pacific Ocean, which normally blow from the east to the west, get weaker than normal, the water around the equator heats up. Because the Pacific makes up so much of the planet's surface area, a change like that can raise the global temperature and cause some of the same kinds of weather caused by climate change—extreme rainfall and flooding in South America, drought in Oceania and Asia, drier winters in the eastern United States and wetter ones near the west coast.

Scientists still struggle to predict exactly when El Niño will happen. Some years experience a strong El Niño, a weak one, or none at all. Other years experience its polar opposite, La Niña, in which the trade winds get stronger than normal. Sometimes the systems come two years apart or seven years apart or longer. And we don't yet know for sure when to expect future ones.

This is yet another kind of feedback loop. Warming temperatures are making a strong El Niño more likely, and the impacts of a strong

water, those structures can fail. Soil and plant roots will soak up some of that water, just like they do during rain, but floods sometimes bring more water than the soil can handle. Floods can drown plants or wash them away, which means there are fewer plants to soak up water during the next flood.

Cities and suburbs have their own sets of problems. Much of their land is covered by buildings, streets, sidewalks, parking lots, and other surfaces that don't absorb water the way soil does. Cities usually have elaborate sewage systems that drain water from the street, but most of those systems were built before these extreme weather patterns set in and aren't always equipped to handle the amount of rain that can hit at once. Many cities also have a lot of buildings with underground components—from basements in houses to full

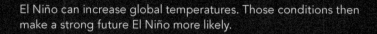
El Niño can increase global temperatures. Those conditions then make a strong future El Niño more likely.

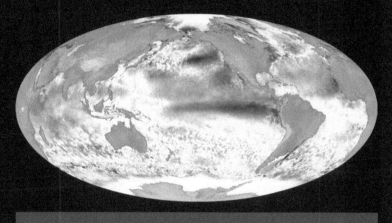
Warming equatorial waters in the Pacific Ocean give rise to the El Niño weather pattern. The darker parts on this satellite map from NOAA show the El Niño from 2016, one of the strongest on record.

garden-level apartments—that are vulnerable to flooding.

Melting snow can also cause similar problems. When more snow falls in a shorter period, more of it piles up, so there is more of it around to turn to water once the weather gets warmer. Many cities clear roads and driveways by plowing snow into huge deposits in a few concentrated areas—which is great, until that snow melts.

Extreme precipitation can also lead to flash floods, or short periods of heavy flooding. While weather services have generally gotten pretty good at predicting weather patterns, flash flooding is harder to predict because of the number of factors at play. That unpredictability makes it hard to prepare for flash floods.

Floods don't need to hit monsoon levels to be dangerous. More Americans die in flash floods than other types and more die from flash flooding than from tornadoes or hurricanes. Running water merely a foot (30 cm) deep can carry a car away, and driving into high waters is one of the most common causes of drowning in a flash flood. Even lower levels of water can knock a person over or pour into a home. Plus, floodwater carries whatever it picks up along the way as it moves, including pollution, human waste, and other things that can hurt you or make you sick. Like most water, it also carries impurities that conduct electricity, so hazards like a downed power line or an underground cable can turn deadlier in a flood.

Even flash floods that don't turn deadly can cause problems. In the summer of 2019, the New York City subway flooded during a particularly bad rainstorm. Water poured in from the street level, and delayed commuters shared videos of the stations; it looked as if it was raining inside. Debris that had built up in the city's storm drains combined with the unusually heavy rain to create this disastrous flooding.

Storm Fronts

The country of Antigua and Barbuda sits where the Caribbean Sea and the Atlantic Ocean meet. The nation consists of the two main islands that give it its name, plus a handful of smaller ones (some with memorable names like Great Bird and Prickly Pear). Christopher Columbus documented the islands in 1493 during his second expedition to the Americas, and there is evidence that people have lived there for more than five thousand years.

In 2017, however, the island of Barbuda was empty of people for the first time in at least three centuries. The government evacuated the entire island during Hurricane Irma. The Category 5 storm, with wind speeds that topped 185 miles (297 km) per hour, started just off the west coast of Africa and caused havoc throughout the Caribbean before hitting the southeastern United States. In Barbuda, it destroyed or damaged 95 percent of the island's buildings. Two years later, the island had rebuilt only a small portion of them and only about 60 percent of its people had moved back. The prime minister, Gaston Browne, publicly blamed not just climate change, but also the industrial countries and corporations causing it, for what the hurricane did to his nation. "They are literally placing our civilization at risk," he said in September 2019. "And that is why we have to be very vocal about climate change."

He's not wrong. The conditions that cause more rainfall and more flooding are doing the same thing with hurricanes—for the same reasons. Plus, hurricanes themselves bring a lot of rainfall and flooding with them, making them (pardon the pun) the perfect storm of extreme weather.

Some hurricanes are large enough to span several states at once. Hurricane Isabel, which made landfall in the southern United States in 2003, caused widespread flooding, downed power lines, beach erosion, and several deaths. Intense storms such as Isabel have only increased in frequency since that time.

Warm ocean water is what gives hurricanes their fuel. It both causes them and makes them stronger by providing them with more water. During hurricane season (June through November in the Northern Hemisphere and November to April in the Southern Hemisphere), the ocean is warm enough that water evaporates and forms clouds. Several rounds of evaporation in a short time can lead to dangerous storms.

As the warm air rises, now packed with moisture, the air that replaces it also heats up and evaporates, and the storm gets bigger. Eventually, the process produces a huge batch of rain clouds that spins as it moves, a motion caused by Earth rotating on its axis. The spinning storm can often be bigger than multiple US states. Hurricane Sandy was about one thousand miles (1,609 km) across, about the distance between San Francisco, California, and Denver, Colorado. There are five categories for these storms, based on the speed of their winds, from 74–95 miles (119–153 km) per hour for a Category 1 storm to 157 miles (252.7 km) per hour or higher for a Category 5. Meteorologists name hurricanes to help inform the public and to avoid confusion when multiple storms are swirling around at the same time. Storms that never grow to hurricane size and strength may receive names just like hurricanes do, and they can still be pretty destructive in their own right. Storms often change categories, usually starting smaller, building up to a peak, and then winding down. Hurricanes are actually at their weakest when they pass over land or over cold water. By that point, they're running out of fuel and dying. Of course, that's also when they can do the most damage because they're coming in contact with the places where people live.

Over the past century, ocean temperature has gone up between 1°F and 3°F (0.5°C and 1.7°C). And while hurricanes were a problem before—there are records of a 1605 storm killing more than one thousand people in Nicaragua—warmer oceans are one reason why they've become larger and more frequent. Rising sea levels (see chapter 4) mean there's more water in the ocean to feed storms. The increases in the amount and temperature of water in the oceans have given rise to a level of hurricane activity previously unknown in recorded history.

In August 2017, Hurricane Harvey, a storm that originated in the Gulf of Mexico before passing over Texas and other states along the Gulf Coast, wiped out more than two hundred thousand homes and businesses. The hurricane forced tens of thousands of people to move, killed more than eighty people, and caused more than

$126 billion in damage. At the time, the US Geological Survey—the federal agency that scientifically studies natural hazards—called it "the most significant rainfall event in United States history in scope and rainfall totals since rainfall records began during the 1880s." And Harvey was "only" a Category 4 storm, with winds of 130 miles (209 km) per hour.

Two Category 5 hurricanes hit the Caribbean and the United States within a month of Harvey. First, the Barbuda-wrecking Irma formed in the Atlantic Ocean near the west coast of Africa on August 30, the same day Harvey made landfall in Louisiana. Irma lasted until September 11, and it hit a whole series of Caribbean islands, including the Virgin Islands, Saint Martin, Turks and Caicos, and

SAME STORM, DIFFERENT NAME

Swirling storm systems with high wind speeds that originate in tropical waters have a few names. You might hear them called cyclones, hurricanes, or typhoons. This can be needlessly confusing because they are all terms for the same kind of storm. The only difference is where the storm forms.

A storm with winds less than 74 miles (119 km) per hour is usually called a tropical storm. Once the winds hit that speed, the storm becomes a tropical cyclone.

At this point, location kicks in. Storms are hurricanes if they form in the Atlantic Ocean or the Northeast Pacific. In the Northwest Pacific (including near Japan, the Philippines, and China), they're called typhoons, while they're simply cyclones in the South Pacific or the southern Indian Ocean (including near India, Australia, and New Zealand).

Because of the way Earth rotates on its axis, these kinds of storms spin clockwise in the Southern Hemisphere but counterclockwise in the Northern Hemisphere. This is called the Coriolis effect (which, contrary to urban legends, does not make toilets in each hemisphere drain in the opposite direction).

Cuba, before hitting Florida and Georgia on the US mainland. In total, the storm killed at least 129 people. In Florida alone, more than six million people needed to evacuate their homes. Irma held its top wind speed of 185 miles (297.7 km) per hour for more than a day and a half and grew as big as 650 miles (1,046 km) across.

Although at the time Irma was the largest Atlantic hurricane in recorded history, at least it wasn't the most destructive. (Irma was only the fifth-most expensive in terms of property damage). But that had more to do with which major cities Irma missed—and sheer luck—than anything else.

A few weeks later on September 16, another Category 5 hurricane named Maria formed, and it caused another round of destruction. Maria hit wind speeds of 175 miles (282 km) per hour. While not as big as Irma, it was deadlier because Puerto Rico was right in its path. The storm was bigger than the entire island. It killed more than three thousand people just in Puerto Rico. The entire island lost power for days and parts of it for more than a month, and the storm caused huge mudslides that brought a second round of damage. Puerto Rico wasn't alone. The Caribbean island nation of Dominica also suffered terribly. More than 85 percent of Dominican homes were damaged, and more than two-thirds of the island's people needed to find new ones.

This all happened in one hurricane season in one part of the world. And these were just the worst storms. There were ten hurricanes and seventeen total named storms that season in the Atlantic Ocean.

That's just what the hurricanes themselves did; there were also some troubling domino effects. Dominica lost most of its agriculture in just a few days during Maria, eliminating both food and income for people who also needed to find a way to rebuild their homes. The flooding ruined the island nation's sewage system, leading to the spread of waterborne diseases. The dust and debris from destroyed buildings filled the air for days and caused respiratory problems. The country's major industry before the hurricane had been tourism, but few tourists chose to visit during the recovery. Two years later, many of the people who lost their homes were still living in shelters.

If the city of New Orleans, Louisiana, which still hasn't entirely recovered from the 2005 Hurricane Katrina, is any indication, then Puerto Rico, Barbuda, Dominica, and other places devastated during the 2017 hurricane season could take years to fully rebuild. And the frequency and strength of hurricanes are only going to grow as the climate warms, increasing the risk that these islands could suffer another major storm soon.

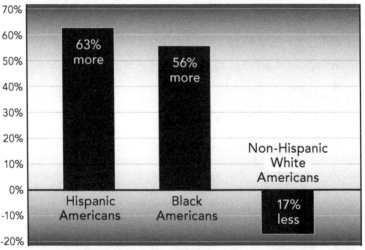

Pollution Exposure by Population, 2003–2015

Source: Christopher W. Tessum et al., "Inequity in Consumption of Goods and Services Adds to Racial-Ethnic Disparities in Air Pollution Exposure, " Proceedings of the National Academy of Sciences, March 2019, retrieved from https://www.pnas.org/content/116/13/6001/tab-figures-data.

In the United States, pollution, climate change, and other environmental issues disproportionately impact people of color. As the data above shows, white Americans actually became less exposed to pollution from 2003 to 2015, while Black and Hispanic Americans were exposed to more.

While 2017 was a record year, 2018 marked the third above-average Atlantic hurricane season in a row, with eight hurricanes and seven other named storms causing more than $50 billion in damage. Although that year's storms weren't record breakers, that wasn't much consolation to people in Florida or North Carolina. And the same year, the other side of the world endured its own cyclone season and more than $30 billion in damage.

You don't even need to live near a coast to see some effects from these kinds of storms. When Hurricane Sandy battered the east coast of the United States in October 2012, its high winds also caused

DOLLARS AND SENSE

The human cost of extreme weather is clearly more important, but it's worth mentioning that the financial costs are also very high.

In 2017, for example, extreme storms in the United States caused more than $300 billion in damage, according to NOAA, with hurricanes alone causing about $265 billion. That broke the US record from 2005, the year when Hurricane Katrina hammered New Orleans and the Gulf Coast. Sixteen extreme weather events in the United States that year did at least $1 billion in damage each.

The insurance industry has understood the risks of climate change for a long time, and companies have conducted their own studies to model its future impacts. After all, when storms destroy property on which the owners have insurance, those companies need to pay up. According to NOAA, hurricane damage to insured property cost the industry more than $515 billion from 1986 to 2015.

Figuring out risk is the key to the insurance business, so companies have already started using climate models to try predicting how much they'll spend on claims—and how much they'll need to charge in premiums to still make a profit. If you live in an area where storms are becoming more frequent, whoever owns the building where you live (whether it's you, your family, or your landlord) is going to wind up paying a whole lot more for property insurance.

huge, near-record waves as far west as the Chicago shores of Lake Michigan—800 miles (1,287 km) away. The walls of water crashing against the Midwestern city created an eerie scene. Winds were as high as 60 miles (97 km) per hour, and waves reached as high as 25 feet (7.6 m). Street flooding and beach erosion played a role in eliminating one of the city's northernmost beaches.

The changes to the jet streams we discussed earlier have been causing hurricanes to slow down, remain longer in one place, and do more damage during those stretched-out visits. Over the past seventy years, these kinds of storms have slowed down by about 10 percent already, giving them more time to affect an area. Of course, the mix of high winds and flooding rain means any hurricane (or even an unnamed tropical storm) is going to cause a lot of damage wherever it hits the shore.

Hurricanes and tropical storms are probably going to get worse unless we manage to slow down or somehow stop climate change, and damage from storms will keep increasing. Charities, governments, and individuals have donated millions of dollars to help rebuild places hit by storms, but bigger and more frequent storms will result in higher costs and more locations competing for those funds. People die in these storms every year, and the indirect impacts such as poverty, homelessness, and disease raise those death tolls.

In other words, warming weather can drive impacts as seemingly minor as rainier days in some places or as major as terrifying storms in others.

CHAPTER 3
FIRE AND OTHER ALARMS

Many of the wars in this century were about oil,
but wars of the next century will be over water.

—Ismail Serageldin, then vice president of the World Bank, in 1995

IN MARCH 2019, the state of California made an important announcement. After eight long years, it declared its record-breaking drought over.

Since the drought began in 2011, at least thirty-five million Californians experienced drought conditions where they lived, and the state had at least some drought every week for 376 straight weeks. California has more people than any other US state (about 39.75 million people, or more than 12 percent of the country) and the biggest economy of any US state (if it were an independent country, it would be the world's fifth-largest economic power, after Germany and before the United Kingdom). California also produces a lot of food for the rest of the country (and to sell overseas), with about 12 percent of US agricultural output even after years of poor growing conditions. A drought in California has a substantial national and even international impact.

"This drought emergency is over, but the next drought could be right around the corner," then California governor Jerry Brown said in 2017 when announcing an end to the worst of the state's emergency conditions. "Conservation must remain a way of life." Under Governor Brown, the state passed numerous laws to conserve water. Some were as simple as requiring restaurants to no longer automatically serve or refill water unless a customer specifically requested it or banning the watering of lawns during the first two days after it rained. Others were stricter, like the five-figure fines for cities and municipalities that failed to hit aggressive water-conservation targets.

Because climate change will make drought an ongoing problem, the state is keeping several of its water-usage restrictions in place. And even though the state's drought emergency technically ended, things are still dire. The land has remained pretty dry, and dry land easily catches fire. Once a dry area starts to burn, it's easy for the fire to spread.

That has been happening in California, over and over, for the past decade or so. Wildfire season is now an annual concern. And according to the California Department of Forestry and Fire Prevention, climate change is responsible for making wildfire season nearly three months longer than it used to be. A 2016 study by the National Academy of Sciences cited climate change as responsible for causing more than 10.3 million additional acres (4.2 million ha) of forest fires between 1984 and 2015 than there would have been without the impacts of climate change. Wildfires in California now burn an area eight hundred times bigger than they did in 1972.

Wildfires are yet another example of an existing natural phenomenon made more extreme because of what humans have done to the climate over the past century. And they're another impact we're already seeing in the present day.

The last official year of that eight-year drought, 2018, was the most destructive in California history when it came to wildfires. In just that one year, the state needed to deal with eight thousand fires. The blazes killed more than one hundred people, with eighty-five killed over seventeen days by one fire in particular—the Camp Fire, named for its place of origin on Camp Creek Road in the Sierra Nevada foothills, near the state capital of Sacramento. At the time, the Camp Fire was the worst fire in state history. While the Camp Fire was burning, another massive fire near Los Angeles caused more than a quarter of a million people to evacuate their homes, killed

THE GRASS ISN'T GREENER IN LA

Watch almost any movie about Los Angeles in its glory days and you'll see lush green lawns and palm trees lining the streets where some aspiring actor is about to catch their big break. One of the many ways people in parts of California dealt with the state's years of drought and conserved water was to get rid of many of those lawns.

With fighting the drought a priority, the government water districts in parts of the state offered financial incentives for people and businesses to replace their grass with less-thirsty options. In just a few years, it's become a lot more common to see houses with unplanted dirt out front, artificial turf, rock gardens, or plants like cacti and other succulents that can get by on small amounts of water. The EPA has estimated that watering lawns and gardens is responsible for about one-third of the country's water use, making it an obvious place to cut back.

While the city looks different these days, it's also more sustainable and better prepared for the next drought.

three people, and destroyed more than one thousand buildings.

Add California's 2018 fires together and they burned more than 1.8 million acres (728,434 ha)—an area larger than all of Delaware or Rhode Island. The fires also destroyed more than seventeen thousand homes and seven hundred businesses, and hundreds of thousands of people needed to pack up and flee their homes in a hurry not knowing if they would ever be able to go back. Thousands of animals living in those burning forests weren't able to get away in time. Then, there was the economic damage: the money and resources the state and cities needed to spend to contain, and eventually put out, the fires; the lost business from people not being able to work; and the time and money it cost to rebuild. Estimates of the 2018 damage go as high as $400 billion, about 2 percent of the entire United States' gross domestic product for the year.

Grass lawns require enormous amounts of water to maintain. Some estimates place the usage in the United States at 9 billion gallons (34 billion L) a day nationwide. In California, many have replaced their grass lawns with native drought-resisitant vegetation, including cacti and other succulents, in an effort to conserve water.

That year was particularly bad, but every year of the drought was extremely destructive. More than 1.3 million acres (526,000 ha) burned just the year before. And this happened after the emergency stage of the drought was declared over. Things could have been worse if the largest fires started a few years earlier.

The record drought officially ending didn't get rid of the fire threat, as California's 2020 wildfire season proved. More than 4.1 million acres (1.7 million ha) burned that summer and fall—more than double 2018's destruction and covering more than 4 percent of the state. That included more than ninety-six hundred incidents, one of which became the first modern "gigafire"—that is, a single fire spanning more than one million acres (400,000 ha). Several things caused individual blazes—from a severe lightning storm in northern California to a pyrotechnic device at a party gone wrong near Los Angeles—but dry conditions helped them spread.

Because such extreme fires have become so frequent, it's easy to forget that they are a relatively recent problem. Even California's track record as one of the leaders when it comes to addressing climate change doesn't protect it. The state passed strong emission limits in 2006 and even stronger ones in 2016. The new ones set a goal for greenhouse gas emissions to be 40 percent below 1990 levels by 2030. And the past decade of fires has given the state's fire-prevention and firefighting personnel plenty of experience. If drought and wildfires can do so much damage to a place as prepared for them and with as many resources as California, that should serve as a serious warning for places that aren't as ready.

The trend of climate change taking existing weather to the extremes, making wet areas wetter and dry areas drier, certainly applied in the case of the California drought. The state had long stretches of abnormally warm weather, along with unusually high atmospheric pressure that pushed the jet stream and the rain it would normally bring north, keeping those warm areas dry. This was partly the fault of a ridge of pressure (which UCLA climate scientist Daniel Swain nicknamed the "Ridiculously Resilient Ridge," or "RRR"), which stayed in place over California's Pacific coast during the winter and blocked storms that should have brought rain and cooled things off. No rain equals longer drought.

Droughts are already one of the most common impacts of climate change. In 2007, the IPCC announced that the amount of land experiencing drought conditions around the world had more than doubled since the 1970s, and drought affects more people than any

ONLY YOU CAN PREVENT THEM

If the area where you live is dealing with a drought, there are some simple, common-sense things you can do to help prevent wildfires. They're good ideas if you live somewhere else too.

- If you smoke, don't toss the remains on the ground, and make sure they're not still burning before you throw them in the trash. Same with matches or disposable lighters. Dry grass, leaves, or garbage can all fuel fire.
- For the same reason, don't shoot off fireworks, even if they're legal where you live. It's not worth the risk. Plus, they're horrible for wildlife (if you've ever seen a cat or dog react to them, you can imagine how other animals feel). Leave it to the professionals.
- Don't drive or idle your car on leaves, grass, or brush. Grass fires can light under your car, which will probably have gasoline in it.
- If you use fire for any reason—such as for campfires, lighting candles, or barbecues—make sure not to leave the fire alone and be sure it's completely out when you're done.
- If you see an unattended fire, don't ignore it. Put it out yourself if you can or call the fire department if it's too big. Don't assume someone else is going to do it.

That all sounds simple—and it is. Don't be the reason a fire turns into a wildfire.

other natural disaster. It also causes other problems, and now we'll talk about them in more detail.

A Dry Heat

The simplest definition of a drought is a period when an area stays dry longer than is normal. Unlike with other natural disasters like a hurricane or tornado, it often isn't obvious at first that a drought is happening. The early days, weeks, or even months of a drought can seem like a normal stretch of dry weather. Like a recession in the economy or a losing streak in sports, a drought is easier to identify after it's been going on for a while.

There are a few reasons why droughts happen. Naturally, rising temperatures themselves help cause droughts. We've talked about how heat makes water evaporate, and that definitely plays a role. So can a lack of precipitation. A wetter area can go without rain for a while without experiencing a drought. But when a place is already dry, a lack of rain allows droughts to start or continue. As we've covered already, climate change is making temperature rises and extended periods without precipitation a lot more common.

Another big factor is the way hotter air is changing weather patterns so that some places are getting more precipitation than normal and others are getting less. When it rains (or snows) less often in places that already don't get very much, it's that much harder for water sources like rivers and lakes to refill. Over time, the amount of water available to evaporate and turn into rain decreases, making things even drier and resulting in even less water for precipitation, and on it goes.

Soil also loses water to evaporation. During droughts, soil can dry out, meaning it isn't able to absorb water as well when the rain finally arrives. Much like how floods can wash away soil, winds can carry dry soil away. (See the sidebar on pages 72–73.) Dry soil and lack of rain both contribute to killing plants that need water to grow.

In extreme cases, drought can cause desertification, which is exactly what it sounds like—an area that used to be fertile land turning into a desert. Human actions like deforestation, poor water use, and unsustainable agriculture—or the combination of these activities with climate-change-induced droughts and other environmental factors—can all cause desertification. While human activity causes a significant amount of desertification, increasing temperatures make the problem worse.

Desertification in turn can cause additional climate change. The Sahara in North Africa, already the world's biggest "hot" desert and the largest desert outside the north and south poles, is now about 10 percent bigger than it was in 1920. Like many other deserts, it expands and contracts somewhat based on the seasons as well, but it's the long-term expansion that is worrying. The Gobi in China and Mongolia has expanded at an even faster rate. Sandstorms from the desert regularly hit urban areas. One May 2017 storm covered more than one million square miles (2.6 million sq. km) of China with desert dust. In both these cases (and many others), deserts have overtaken some of what used to be farmland and have forced people who lived near their borders to relocate. China has planted more than sixty billion trees in a "Great Green Wall" to try to limit the Gobi's expansion, creating the world's largest human-made forest in the process, but the results have been mixed at best.

It's not only land that can dry out. The Aral Sea in central Asia used to be one of the biggest freshwater bodies in the world, covering 26,000 square miles (67,000 sq. km) between Kazakhstan and Uzbekistan. Over the past few decades, it dried out so much—largely because of people taking water from it for irrigation—that it shrunk to 10 percent of its original size by the mid-1990s. In that time, it split into smaller bodies of water. As its fresh water disappeared, the saltwater resting underneath the fresh water increased the Aral's overall salinity and killed the freshwater fish that had thrived there

DUST TO DUST

The Dust Bowl period in the Midwestern United States more than ninety years ago predates the worst of the climate crisis, but it's still a good example of the kind of extreme conditions we should try very hard to avoid.

A series of droughts started in 1930, drying up massive amounts of farmland and killing crops. What made it worse was that farmers in that part of the country had already cleared natural grasslands and trees to plant wheat and other crops and had been producing more than the land could sustain. World War I (1914–1918) and the damage it caused led to an increase in demand in Europe for imported food, and American farmers supplied it. Overfarming damaged the soil and prevented crops from spreading the deep roots needed to hold soil in place. When high winds came to the plains, they turned the loose soil into huge dust clouds that blew over large parts of the country.

A serious drought in 1934 affected almost three-quarters of the United States. One storm that year featured clouds of sand and dust that blocked the sun for two days and carried dust more than 2,000 miles (3,219 km), as far east as New York City. Ground temperatures felt colder than normal because the dust blocked sunlight, and the same lack of sunlight caused even more crops to fail.

Because it occurred during the Great Depression, the Dust Bowl made an already-damaged US economy (and Canadian economy, since plains there were also hit) worse. The large amount of farmland affected caused food supplies all over the country and exports to other nations to suffer. More than two million people had to leave their homes, many heading west in search of work.

for ages. What used to be fishing villages on the shores of the Aral Sea instead became desert ghost towns.

The region saw winds blow sand and dust over wide areas. Salt from the dried sea also blew around central Asia, killing plants that had never interacted with salt before. The dust also carried pesticides, fertilizers, and other dangerous chemicals. Dust storms from the Aral Sea area have been linked to the area's higher-than-normal rates of infant mortality, cancer, and and other diseases. Because of

By the end of the 1930s, some parts of the United States had lost 75 percent of their topsoil. Under president Franklin Delano Roosevelt's "New Deal," the federal government took action to fix the problem, from planting trees to setting prices to enforcing farming techniques that would keep soil from eroding. It took a while but helped tremendously, and the return of normal rain in 1939 finally ended the Dust Bowl conditions after nearly a decade.

A later NASA study, based on tree-ring data and other sources, found that the drought that caused the Dust Bowl was the worst the planet had seen since 1000 CE, and affected seven times larger an area than any North American drought in more than one thousand years. The good news is a worst-case drought like the Dust Bowl is unlikely to occur again soon, and we're more prepared now, though future droughts don't need to be nearly that extreme to still do major damage.

A dust storm looms over a Colorado town during the Dust Bowl.

the discovery that the Soviet Union had used former islands in the Aral Sea, now connected by land, for Cold War-era experiments with anthrax and bubonic plague (as in the black death, the same kind of plague that wiped out much of Europe in the Middle Ages), cleanup crews were needed to prevent that stuff from spreading too.

Thankfully, through a long and expensive process, including more than $80 million in aid from the World Bank, Kazakhstan has been able to bring back some of the northern part of the Aral

Sea. Creating dams has helped build up water supplies, and fish are now able to survive again in that part of the Aral. It will probably never get as big as it used to be, and the southern part near Uzbekistan is still dried out. However, those restoration efforts have managed to reduce some of the desertification effects in the region.

Despite how bleak this scenario and so many others may seem, the Aral situation demonstrates that it's still possible to improve things. Of course, preventing these kinds of environmental problems from getting worse—or from even happening in the first place—is a better approach than trying to fix them afterward.

A 2013 United Nations report found that Earth had lost about one-third of its arable land (land capable of growing crops) in just the past forty years. Even worse, a series of 2018 studies found that even if the world were able to limit greenhouse gas emissions enough that temperatures only hit 3.6°F (2°C) above preindustrial levels, about 30 percent of the total land on Earth could still dry out. The United Nations has predicted that more than 1.2 billion people—and perhaps as many as 2 billion—will be impacted by growing deserts before the end of the twenty-first century. More than half its member countries have joined in efforts to prevent desertification and replant affected areas.

Thirsting for Relief

Like the desert examples above show, people play a role in droughts beyond our general impact on the climate. On the flip side, droughts also have other effects on us and make life harder.

We need water to grow food and to survive. Most people can only go about three or four days without water before they die of dehydration. During droughts, people rely on groundwater, reservoirs (either natural or human made), snowpack, and other sources of saved-up water. In the short term, that saves lives. But in the long term, it can cause problems when rainfall doesn't replenish those sources and there's less water available for next time.

For example, people in much of the world still get their water from wells. They tap into groundwater, the underground water deposits that build up after passing through gaps in soil, rock, and other materials. About half of the drinking water in the United States comes from aquifers—groundwater deposits—and they're the main source of water for irrigating crops. So, when they dry out, both the food and water supplies suffer.

While 71 percent of Earth's surface area is covered with water, most of that is saltwater; only about 3 percent is fresh water. Many places around the world (from the United States to Saudi Arabia to Kiribati) convert saltwater from oceans and seas into drinking water by removing salt and other minerals, a process called desalination. Somewhere between fifteen thousand and twenty thousand desalination plants operate around the world. Most of the largest are in the Middle East, where fresh water is scarce. But desalination plants require a lot of energy to operate, producing a lot of greenhouse gases in the process. Plus building and operating the plants is expensive. Eliminating saltwater is also not great for the oceans and the sea life that lives there, so desalination solves one problem while creating others.

Even with all the warning signs science has identified and all the steps we've taken to prepare for the worst, water can still be hard to come by during droughts. Cape Town, one of the three capitals of South Africa, had its own recent water crisis. In January 2018, after four years of the worst drought in more than a century, local officials announced that Cape Town was about to run out of water. They determined that the city had only ninety days to fix things before all of the municipal water ran out and set a specific date as "Day Zero," on which there would be no water at all. The story got attention worldwide, partly because a modern city with four million people, quality infrastructure, a strong economy (responsible for about 10 percent of its country's gross domestic product), and a thriving tourism industry didn't fit the stereotype of what an area about to run out of water looks like.

It took a huge government effort, but Cape Town dealt with the crisis through a combination of reducing water use, changing farming practices, and sourcing new ways to get water. The government issued strict water rations, with residents allowed to use only 13.2 gallons (50 L) of water per day. That might sound like a lot, but many people use much more in their daily activities. The EPA estimates that the average American shower uses about 2.5 gallons (9.5 L) of water per minute, so a ten-minute shower uses about 25 gallons (94.6 L)—almost double what Cape Town residents were allowed for an entire day. If you drank the recommended eight glasses of water a day, you'd consume about half a gallon (1.9 L). Add showering and drinking to cooking, cleaning, or a myriad of other everyday uses of water that are easy to take for granted. Even flushing a toilet can use 2.4 gallons (9 L), which explains why Cape Town businesses put up signs in public bathrooms encouraging people that "if it's yellow, let it mellow," plus similar signs asking them to opt for hand sanitizer instead of using water to wash their hands. In all, Cape Town limited individual water use to only a small fraction of the limit California needed to put in place during its own terrible drought.

Along with those restrictions, Cape Town fined people who violated the limits, and published household water usage so people could see whether their neighbors were following the rules—a "name and shame" strategy that relied heavily on peer pressure. Use of underground springs was regulated, with people waiting in long lines to get set amounts of water. Farms needed to change their water usage too, and the city contracted with new sources to temporarily bring in extra supply.

After four years of drought, Cape Town finally got enough rain in June 2018 to climb out of a crisis state, and "Day Zero" didn't arrive. Though the city avoided the worst-case scenario, it still had to cope with a moderate drought. A year later, the city's dams still held only about half their capacity of water, and water restrictions were still in effect.

Chennai, India, was not quite as lucky. A city of more than ten million people, Chennai saw its major reservoirs fall to their lowest levels in at least seventy years in 2019. Some went completely dry, with aerial photos showing only mud and cracked ground where huge lakes normally sat. The city's groundwater was already under stress, falling by 85 percent in a decade, so it was pretty much tapped out too.

Climate change and drought were factors, of course, but so was lack of conservation. For years, water use was poorly regulated. In a case of poor planning, the city expanded by building on top of wetlands, and Chennai's population grew too quickly for resources to keep up. The city was basically balancing on the edge of a cliff, and the annual monsoon (see page 52) arriving later than normal pushed it over the edge. Without the rain that monsoon would have brought, the water supply wasn't replenished, and it was already too late to try some of the policy options that worked in Cape Town.

The city wound up relying on water tankers that carried supplies from bore wells—deep dugouts in the countryside outside Chennai that still had some caught rainwater from the last rainy season. Water became such a valuable resource that there were news reports of tankers being hijacked at gunpoint and people fighting each other over water. Income inequality is a bigger problem in India than in many countries, and water became something the wealthy could buy from private companies while others couldn't afford the crucial resource.

In June 2019, the Indian government predicted that because the country's population is growing dramatically at the same time that climate change is making water supplies precarious, India's water demand could be twice as high as its supply in less than a decade. The Chennai crisis has spurred some future action, including building more desalination plants and updating India's rain-catchment systems. Still, as one official told the BBC during the 2019 crisis, "The destruction has just begun. If the rain fails us this year too, we are totally destroyed."

By 2018, other major world cities including Mexico City, Melbourne (Australia), São Paulo (Brazil), and Jakarta (Indonesia) faced the possibility of running out of water and took steps to limit use and increase supply. Some of those situations improved once droughts ended. But for a lot of the world, the casual use of water is already a thing of the past.

As climate change raises the heat and makes dry places drier for longer, more cities are going to join the list. In August 2019, the World Resources Institute reported that about one-quarter of the world population already faces an imminent water crisis.

The experience of these cities reinforces another important point. Scientists have been warning for decades about the kinds of droughts that California, South Africa, India, and many other places have dealt with lately. As we discussed in the first chapter, too many people still think of climate change as a future problem when it's actually a present one. They don't always get it until there's an immediate crisis, something dramatic they can't ignore—such as a wildfire.

A Burning Problem

After Australia's 2019 winter, the one with the unusual snows, parts of the country started burning. By September, much of the landmass was on fire. The last of the fire wasn't brought under control until February 2020, when a week of heavy rain put out enough of it that firefighters could contain the rest. In other words, a whole continent burned for most of a year.

Wildfires, or bushfires as people there call them, are nothing new in Australia. Indigenous Australians used limited fires as a technique to clear land thousands of years ago. The landmass tends to be hot in general, and bushfires are an annual spring and summer event in the dry inner part of the country. (In the Southern Hemisphere, spring and summer run from September to March.)

Fires can spread quickly in Australia because native trees like eucalyptus are very flammable. When the oil in a eucalyptus tree gets hot, the tree releases vapor. (There's often a visible bluish mist from that gas around trees in a eucalyptus forest.) The oil catches fire easily and spreads it too. Along with that, eucalyptus trees shed leaves and bark that form piles on the ground beneath them, which is basically perfect kindling.

Wildfires have increased in both frequency and intensity in drought-prone areas. One way to prevent out-of-control wildfires is to conduct controlled burns, which are intentional fires set to help clear the land of particularly combustible materials, such as dead plants. Wildfires and controlled burns are necessary for ecosystems to remain stable and healthy.

So, when drought conditions in Australia get bad, fires start easily and become very destructive very quickly. Although bushfires have happened in Australia for a long time and the plants and animals there have generally adapted to survive in an environment where they happen, rising global temperatures have thrown things off balance. Bushfires are more intense and spread wider than ever before.

By September 2019, thanks to an extended drought and high temperatures, firefighters around the country had to deal with more than one hundred individual fires at the same time. The northeastern state of Queensland already had fifty-seven bushfires raging at that point—the earliest in the season that the state had major fires since it started keeping records in the late 1800s.

Despite the best efforts of the country's various fire departments, more than fifty of those fires hadn't been contained by mid-December, and roughly another fifty were contained but still burning. A fire near Gospers Mountain in the state of New South Wales was the biggest individual blaze in the country's recorded history. It burned an area more than seven times the size of Singapore, including areas that had been free of fire for at least half a century—another indicator that this wasn't a normal bushfire.

In total, more than 40,000 square miles (103,600 sq. km) throughout the country burned in just one fire season, and some estimates are even higher. That's still more than one-fifth of the whole continent's forests, and more than a dozen times as much land as was affected by the California wildfires. Thousands of homes were damaged or destroyed, and at least thirty-three people died. Australia's unique wildlife suffered horribly, with early estimates predicting as many as half a billion to one billion animals killed by fire and smoke and several species at risk of extinction. (See the sidebar on pages 82–83.)

The air in major cities along the coast, such as Sydney and Brisbane, filled with dangerous levels of smoke. That smoke was so bad that glaciers on the west coast of New Zealand—more than 2,500 miles (4,023 km) away—turned pink and eventually brown from the ash and dust blown across the Tasman Sea. Even after most of the fires had been extinguished, their ash and other debris contaminated Australian rivers and other waterways, killing even more wildlife and ruining an important source of fresh water. These scenes looked like something out of an apocalypse movie and could very well happen again.

On one Tuesday in December, the average high temperature across Australia was 105°F (40.6°C), breaking a national record. The record only lasted one day before Wednesday's average hit 107.4°F (41.9°C). Temperatures in individual parts of Australia get hotter than that pretty often, but that was the highest nationwide average. The western city of Perth had three consecutive days above 104°F (40°C), which had never happened in December before. A 2018 study by the Australian government's Bureau of Meteorology found the country had already warmed a full degree Celsius over the past century, meaning Australia was more than halfway to the two-degree mark we discussed earlier. And that was before the record bushfires got going.

Coincidentally, the record-hot day came the same week that the 2019 United Nations Framework Convention on Climate Change, an international summit of nearly two hundred countries to address climate-change policy, ended without an agreement on cutting carbon emissions and tabled the issue until its 2020 meeting. While that 2020 meeting didn't get a deal either, the convention at least doubled the number of entities (states, cities, investors, and businesses) committing to reach net-zero emissions by 2050, with Facebook, Ford Motor Company, and the state of New South Wales among the big names joining the list.

The Australian federal government wasn't an innocent bystander in all this either. Prime minister Scott Morrison had gotten rid of

A SYMBOL OF WHAT'S AT STAKE

As bushfires destroyed much of Australia in 2019 and 2020, they also killed untold numbers of Australian animals. Koalas, in particular, became an international symbol of the fire, as thousands of the beloved animals perished or were seriously burned. While they can move faster than you might think, koalas are still fairly slow when it comes to outrunning danger. Their natural defense is to climb the eucalyptus trees where they eat and sleep. Normally, they can climb higher than a fire will spread upward. That had worked for millions of years. These fires, however, burned at a higher-than-normal intensity. The fires could reach the treetops, and eucalyptus oil made some trees ignite completely. Koalas weren't safe anywhere the fire spread.

In one memorable video, an Australian woman named Toni Doherty rushed into a blaze near Port Macquarie to save an injured, crying koala from a burning tree, wrapping her shirt around him and running back to the car in time to get away from the fire. Doherty's heroism went viral, though the koala (whom she named Lewis after one of her grandchildren) was so badly burned that—despite pain medication and a breathing apparatus—vets needed to euthanize him just one week later. Other koalas with some or all of their fur burned off became living warnings about wildfires. Zoos and other organizations made special efforts to take care of them and other injured animals.

Koalas were already considered vulnerable (not yet endangered, but the level just one step from it) before the bushfires grew out

many emission-cutting policies implemented by his predecessors, and Australians rank among the highest per-person emitters of greenhouse gases. Because the country still relies on coal rather than green energy and is the world's leading coal exporter, Australia contributes more to global emissions than you might expect for a country of only about twenty-five million people.

Australia's bushfires aren't just a consequence of climate change; they're also a contributing factor. According to NASA, in just their first three months, the 2019 fires had already emitted more than of control, mostly because of habitat loss and a recent disease outbreak. While there were some initial fears that the 2019 fires put the marsupials at risk of immediate extinction, that isn't yet the case. But koala populations in some parts of the country may never fully recover, and fires have erased more of their already scarce habitat. Unfortunately, the worst could still occur if fires this intense become normal.

A caretaker holds a rescued koala after a bushfire. Some environmental groups predict koalas will be extinct by 2050 unless action is taken to stop the destruction of their habitat.

250 million tons (226.8 million t) of carbon into the atmosphere, about half of what Australia usually produces in a year. As the fires continued to burn more than a year later and with the loss of trees and plants that would normally take in some of the carbon from the atmosphere, these bushfires helped make future ones longer and more intense.

The Australian crisis also came almost immediately after a set of forest fires in the Amazon rain forest in Brazil gained worldwide attention. These fires had a very specific human cause—

deforestation. From January to September 2019, peak logging season, people had cut down about 85 percent more trees than they had the previous year (and it's not as if that wasn't already a dangerous amount of logging). The result? About 30 percent more fire than the year before and more than seventy thousand fires in the rain forest in just that short span. Some of those were set by people deliberately trying to burn off tree trunks and leftover wood to clear the land for farms and ranches. Others started because of high temperatures and drought. At one point, the equivalent of one and a half soccer fields was going up in flames every minute of the day.

With logging, agriculture, and other businesses clearing forest, an estimated 15 to 20 percent of the Amazon is at risk of deforestation. While the situation is serious, a large amount of the Amazon rain forest is federally protected in Brazil and should not be at immediate risk unless local laws change. But the government leaders the country elected in late 2018 have encouraged changing those laws to open the Amazon to more farming and other business interests.

A burning Amazon has additional consequences when it comes to climate change. As we've discussed, plants take in carbon dioxide and produce oxygen. Because the Amazon is by far the world's biggest rain forest, spanning 2.1 million square miles (5.5 million sq. km), more of the carbon-intake process happens there than in any other spot on the globe. We'll talk about this in more detail in chapter 7, but carbon capture is one of our best hopes for reducing the amount of carbon in the atmosphere and—hopefully—preventing, or at least mitigating, some of the worst impacts of climate change. Losing so much of such an important carbon-capturing area means even more future work will need to be done to make up that difference. If more of the forest disappears, more carbon capture would need to come from somewhere else. The Amazon rain forest is important for many other reasons as well, including that it is home to a few hundred

Indigenous tribes and also to a diverse group of plants and animals (some found nowhere else).

Then there's the carbon already trapped. As the huge emission levels of the Australian fires demonstrated, burning forests release massive amounts of the carbon they currently hold, and the Amazon holds around 17 percent of Earth's trapped carbon. Scientists have noted that wildfires alone could release more than a billion tons (900 million t) of carbon from the Amazon. Fires there both add to the emissions in the atmosphere and eliminate a very valuable way

Deforestation of Borneo

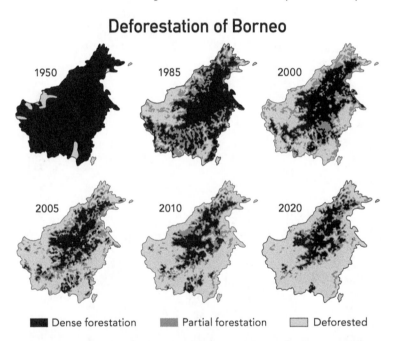

1950   1985   2000

2005   2010   2020

■■ Dense forestation   ■ Partial forestation   □ Deforested

Source: https://www.researchgate.net/figure/Extent-of-deforestation-in-Borneo-1950-2005-and
-Projection-Towards-2020-Source-PEACE_fig3_309747458

The island of Borneo, home to the Borneo orangutan, has suffered tremendous deforestation as farmers replace the diverse rain forest with palm trees. The trees produce palm oil, a common ingredient in many foods, from candy to livestock feed. As demand for palm oil increased, the Borneo orangutan's habitat and population dwindled considerably.

of subtracting them. The Amazon also plays a big role in the planet's water cycle. Rain forests produce quite a lot of their own water (in the Amazon it's around half). The rain there is captured by soil, tree roots, and other systems, and that water evaporates and turns back into rain. A lack of trees means less water is absorbed, so less rain falls, and potentially more drought and more fire will occur.

As you can see from all these examples, droughts aren't a future problem, but one that's already happening and will keep getting

KEEPING IT REAL

As the many examples in this book and any number of reliable news sources can tell you, there's plenty to actually worry about when it comes to climate change and Earth's environment. Still, in the age of social media, it's possible for even people with good intentions to spread inaccurate information that can make an already terrible situation sound even worse.

During the Amazon fires, many people (including multiple countries' presidents) shared their sadness over the destruction while calling the rain forest the "lungs of the Earth" and a source of "20 percent" of the planet's oxygen. The rain forest does a lot of amazing things, but that part isn't entirely true. The Amazon rain forest produces only a fraction of that much oxygen, and the forest environment uses up much of the oxygen that it produces. Still, that 20-percent number has been referenced for a long time, even though it's misleading. National Geographic says it goes back decades and suggests it might be a misreading of a real stat: that about 20 percent of the oxygen produced by photosynthesis on land comes from the Amazon. Still very important, just different.

News organizations like the New York Times, National Geographic, and CNN jumped in to explain things, without downplaying the destruction caused by the fires. As with any current-events topic, media literacy and fact-checking are always handy skills if you want to better understand what's going on with the environment. The way a well-meaning error went viral is also a good reminder of how easily misinformation can spread.

The Amazon rain forest is threatened by profit-driven agriculture. Huge swaths of forest are cleared daily to make room for farming and ranching, including crops used to feed livestock.

worse. Like the extreme weather we talked about in the last chapter, droughts, water shortages, and fires caused or enhanced by climate change have already affected millions of people. In the next chapter, we'll talk about another problem that's already doing serious damage—something else that needs to be stopped before it's too late.

CHAPTER 4

THE TIDE IS HIGH

Either we stand united and agree to combat climate change, or we will all stumble and fall and condemn humanity to a tragic future.

—Enele Sopoaga, then prime minister of Tuvalu, during the 2015 Paris climate talks

EVERY SPRING, THE country of Tuvalu floods as the annual "king" tides crash over the low-lying coral atolls. Huge surges of water from the Pacific Ocean push over the coasts. Because they come from the ocean, the tides bring saltwater that seeps through the coral land into the country's groundwater and its freshwater lagoon. The saltwater contaminates the limited natural drinking water and kills some of the starchy root vegetables such as taro and pulaka that Tuvaluans have grown as a regular part of their diet for centuries.

On the main atoll of Funafuti (the one discussed in the introduction), the water now goes so far inland that the runway for the country's international airstrip, made of coral and built in the center of the island by US forces during World War II (1939–1945), floods. So do many people's homes and the few roads that run between them. The highest point in all of Tuvalu, on its southernmost island of Niulakita, is only about 15 feet (4.6 m) above sea level. Most of the country's nine islands are less elevated than that. And thanks to climate change, the tides that used to be more of a seasonal disruption have turned into a threat to people's ways of life.

All tides are caused by the force of gravity, with the gravitational pull of the moon and the sun moving the ocean tide back and forth. Tides are normal, and scientists can predict high and low ones pretty accurately. What has changed lately is how much the water in the ocean is rising. And more water moving back and forth as a result of this rise can create higher tides.

As far back as 2000, tides more than 10.5 feet (3.2 m) high flooded Tuvalu and even knocked out the main island's phone service for days. In 2006, the king tides that covered the islands measured as high as 11 feet (3.4 m), and the tides were that high again in 2015. Other recent years haven't broken those records, but they're in the same ballpark and are yet another example of how climate change is amping up an existing phenomenon to an extreme degree.

To put those numbers in perspective, in the United States, a 10-foot (3 m) flood is (rightly) seen as a huge disaster. When Hurricane Katrina flooded and devastated New Orleans in 2005, the worst of it was about 15 feet (4.6 m) of water. With rising sea levels proving a consistent problem, floods that bad are now a regular thing that people on low-lying islands experience. As the EPA put it in 2011, the annual king tides will become the future's everyday tides.

WHAT IS AN ATOLL?

An atoll is a thin coral reef that grew around an underwater volcano. When the volcano erupted, the lava piled up until it broke the surface of the water, forming an island. Corals grew around the edges of the island and formed reefs over thousands or even millions of years.

As time passes, the island erodes and sinks, leaving a shallow ring of land formed by the reefs. This process happens in stages. When the island in the middle has shrunk and a lagoon begins to form around it, the coral surrounding the lagoon is known as a barrier reef. Once that central island completely disappears, leaving only the ring of coral around the lagoon, the reef is considered an atoll. Most lagoons contain fresh water, as the coral barrier at least partly protects the water from the ocean around it, letting rainwater build up over time. Because saltwater is heavier than fresh water, the fresh part sits on top, forming what's known as a water lens, which allows freshwater fish to survive there. The coral rims of the atoll continue to grow, and that process has sustained them for millennia, but sea levels are rising too fast for the growth to keep up. Along with their low elevation, atolls are particularly vulnerable to rising seas because the land itself, being made of coral, is porous.

In 1842, Charles Darwin became the first scientist to publish an accurate theory of how atolls form. He came up with it during his productive voyage aboard the HMS *Beagle*, the same one on which he circumnavigated the globe and formed his theory of evolution by natural selection. It took a few more decades and a series of experiments before other scientists proved him right, but Darwin's work on atolls held up. Funafuti, Tuvalu, was one of the locations where his theory was later tested. The spot used for drilling tests there has a small coral marker commemorating it.

The IPCC made plenty of headlines in 2001 when it predicted the sea level would rise by at least 3.5 inches (9 cm) and as much as 34.6 inches (88 cm) by the end of the twenty-first century. It was a bold and scary prediction based on a combination of models, measurements using tools like tidal gauges and satellites, and

After Hurricane Katrina devastated New Orleans, Louisiana, the flooding was so severe that entire houses were submerged.

past experience. Some critics accused the panel of exaggerating or relying too much on old data. As it turned out, the prediction was not entirely accurate—because it was too careful and conservative. Emissions kept increasing, and the sea level rose with them. By 2007, the IPCC already needed to revise its prediction to at least 11 inches (28 cm) by the end of the century—making the updated best-case scenario three times worse than what the panel had predicted just six years earlier.

The IPCC released yet another update in the fall of 2019, which warned that based on the faster-than-expected melting of the Greenland and Antarctic ice sheets (more on them shortly), sea levels will likely rise at least 3 feet (92 cm) by the end of the century. So, the most likely outcome is now worse than what the 2001 report described as the worst-case scenario. And as the IPCC pointed out in that 2019 study, we might only have until 2030 before we hit the dreaded point of no return at which sea-level rise can't be stopped.

The past has already shown a rising trend. From 1880 to 2019—in other words, from when the Industrial Revolution really got going up to the present day—worldwide sea levels rose by about 8 inches (20 cm). That's an average of about 0.07 inches (1.8 mm) per year. But from 1993 to 2010, the average sea level rose about 0.13 inches (3.3 mm) per year. Even if the sea level continues to increase only by that amount, it would reach more than 11 inches (28 cm) above the already-high 2001 level before this century ends. It is more likely that the rate of sea-level rise will keep increasing, considering the overall warming patterns we discussed earlier.

Scientists predict that sea levels only need to rise about 8 to 16 inches (20 to 40 cm) from where they were at the start of the twenty-first century—a number right in the middle of that expected range—to make Tuvalu completely uninhabitable. (Keep in mind that any place becomes unlivable for humans well before it's permanently underwater.) Tuvalu's loss is probably a matter of when it will happen, rather than if it will. And Tuvalu isn't alone. Other low-lying island countries are in pretty much the same situation.

The archipelago nation of the Maldives is made up of more than one thousand islands, though only about two hundred of them are inhabited. The country, the smallest in Asia, sits southwest of India in the northern Indian Ocean and the Arabian Sea. It is known for its beautiful white-sand beaches that attract tourists from around the world. It is also the least-elevated country in the world. More than 80 percent of the Maldives is less than 3.3 feet (1 m) above sea level.

Two Pacific countries, Kiribati and the Marshall Islands, are also just a little bit higher than the sea level. Kiribati consists of thirty-three islands, but they're spread out across a wide expanse of ocean, spanning about 1,800 miles (2,900 km). More than 80 percent of that total land would be uninhabitable if the sea level rises 2.6 feet (0.8 m). The Marshall Islands include more than 1,200 islands, which form two major chains. Nearly all Marshallese people live near the coasts of those various islands, so rising seas would either push them inland or force them to leave their homeland, which many have already done.

While these four countries are most vulnerable from sea-level rise, many other island nations in the Pacific and elsewhere (from Bahrain in the Persian Gulf to Comoros off the east coast of Africa to the Bahamas not far from the southeastern United States) are also at risk. For more than a decade, island nations have been encouraging international agreements to reduce emissions and fight climate change. Although their populations add little to the global emissions problem, they're already paying the price for the actions of bigger, richer, more industrialized countries.

People have lived on most of these at-risk islands for a long time. In Tuvalu, people came by boat from other Polynesian islands in the 1300s, if not earlier. The first settlers of the Maldives arrived at least two thousand years ago. For Kiribati, settlement started between four thousand and five thousand years ago. The land that forms these small nations has housed people continuously for ages.

This is another reminder that the amount that sea levels are rising is not normal or part of a natural cycle, but is yet another problem brought about by the changing climate. The changes people living on these islands are seeing have only been occurring in the past few decades. James Hansen, the prescient scientist who put global warming on the public map back in the 1980s (see page 34), has called rising seas the most dangerous of all climate-change outcomes and one we're running low on time to solve.

A HARDLY LEVEL SEA

When scientists talk about rising seas, the term *sea level* refers to the mean, or average, global sea level. Because Earth's oceans are technically one connected body of water and the smaller seas are included within the oceans, the term applies to all of that water.

Depending where you are on the planet, the normal local sea level could be higher or lower than the global average, for reasons ranging from regional ocean currents to differences in land elevation.

The tides also affect the sea level. On a typical day, most places get two high and two low tides. (If you want to get technical, it's actually two of each every twenty-four hours and fifty minutes, as a single rotation of the Earth with respect to the moon, or a lunar day, isn't exactly twenty-four hours long.) The difference in sea level at high and low tide can be extreme. For example, the Bay of Fundy in Canada—in the eastern part of the country, between the provinces of New Brunswick and Nova Scotia—routinely sees high tides more than 50 feet (15 m) higher than the low tides in the same place. That's why scientists rely on averages when discussing the subject; they're a much more accurate basis for comparison than the wild swings.

The average global sea level is also a useful measurement for defining the elevation of various places on Earth. The highest natural point on the planet is the summit of Mount Everest in the Himalaya mountains, which is 29,035 feet (8,849 m) above sea level. To put that in perspective, Earth's tallest building, Burj Khalifa in Dubai, United Arab Emirates, is only about 2,717 feet (828 m) tall. Earth's lowest natural point is in the Mariana Trench in the western Pacific Ocean, located about 36,000 feet (10,973 m) below sea level. That's more than twenty-four times the height of the Empire State Building.

Ice in Polar Position

Rising sea levels clearly cause flooding, but this is a separate phenomenon from the kind of flooding extreme rains and hurricanes bring. Because it isn't caused by heavy storms, flooding caused by sea-level rise is sometimes called by deceptively pleasant names such as "sunny day" or "nuisance" flooding. Such flooding has a few causes, but the main one is the melting of the polar ice caps and other sea ice.

The polar ice caps in the Arctic and in Antarctica house most of the world's ice. Because the poles are the coldest places on Earth— they're always tilted away from the sun, so they receive less solar energy and have colder surface temperatures—water at the poles freezes more easily than it does at other latitudes.

Both the Arctic and Antarctic ice caps are millions of years old. They also hold most of the planet's fresh water; as much as 70 to 75 percent of it is frozen in the polar ice caps and other glaciers around the world. The fresh water available for us to drink is just a tiny percentage of that.

The two regions themselves are quite different from each other. Antarctica is one of the Earth's seven continents, a massive landmass of more than 5.4 million square miles (14 million sq. km). It is bigger than Europe or Australia. There, the ice formed on top of and along the perimeter of that giant chunk of land. In the Arctic, on the other hand, much of the ice sits on top of water because there isn't much actual land. Greenland and parts of Canada are closest to the north pole, but no true land sits at the pole itself. The glacier known as the Greenland ice sheet is the closest northern equivalent to the Antarctic ice sheet and covers about four-fifths of Greenland.

The north and south poles have seasons just like anyplace else, which cause natural changes to the temperature and to the amount of ice there. In the summer, the poles get warmer and some of the ice melts. In winter, the poles get colder and some of the water and water vapor there freeze and turn to ice. That means there are times during the year when the amount of polar ice actually goes up. Science deniers will often point that out as if it somehow invalidates the larger problem. What they're ignoring is that the amount of polar ice goes up only if you compare a winter total to the respective summer total. Overall, the ice caps lose more ice in the summer than they regain during the winter, and that's one of the most dangerous impacts of climate change because it can't be undone.

NOT JUST A POLAR PROBLEM

Melting ice at the poles is devastating, of course, but those are not the only places it's happening. As noted earlier, glaciers and other ice deposits around the world have been melting and shrinking as the planet gets hotter, and their disappearance is some of the most visible proof of the changing climate.

For example, Glacier National Park in northern Montana, as its name suggests, contains a number of ice masses. In the mid-1800s, before the Industrial Revolution, the area that became the park (in 1910, during the William Howard Taft administration) had about 150 active glaciers.

But this 1,544-square-mile (4,000 sq. km) park by the Canadian border is losing what made it so special. In 1966, the park was down to sixty-six active glaciers, each with a unique name. By 2015, only twenty-six of them remained, and they had all gotten smaller. According to the US National Park Service, the remaining glaciers have shrunk by an average of 39 percent, with some losing up to 85 percent of their former size. The Park Service has published before-and-after pictures of several of those glaciers, taken in the early twentieth century and again in the past two decades, and the difference couldn't be more dramatic.

At this rate, the park's very name is at risk of becoming misleading.

To be clear, the fact that ice is melting at the poles doesn't mean those areas will look like the beach in summer anytime soon. The sheer amount of ice there (millions of square miles of it) ensures it will still take many years before the poles completely melt. The caps just have much less ice now than there used to be. All you have to do is look at NASA's satellite images of the poles over the past few decades to see the ice doesn't expand as far as it used to in the winter and that it disappears during the summer.

People have studied Arctic ice and its seasonal changes since at least the Viking days more than a thousand years ago. As seafaring people, Vikings needed to know when their ships could safely get through the ice so they could do things like raid Ireland and visit

The Boulder Glacier in Montana's Glacier National Park has lost much of its ice in the past century. On the left is a photograph of the glacier as it appeared in 1932; on the right, a photograph taken from the same location in 2005. More such photo comparisons can be found on Glacier National Park's website. https://www.nps.gov/glac/learn/nature/glacier-repeat-photos.htm.

North America. But modern and international efforts to measure the Arctic ice with consistent results date to the 1950s. Similar Antarctic data followed several years later, finding a lot of melting ice.

While average temperatures have gone up everywhere around the world, the Arctic region has been warming about twice as fast as the rest of the planet. Antarctica is also warming faster than the rest of the planet, but not quite as fast as the Arctic. This is partly because of rising air temperatures, but also the fact that water absorbs energy from the sun. The same warming pattern seen on land and in the air is happening in the world's oceans. But water absorbs much more of that energy than land does. Since 1970, the oceans have absorbed about 90 percent of the planet's extra heat, with the land and atmosphere

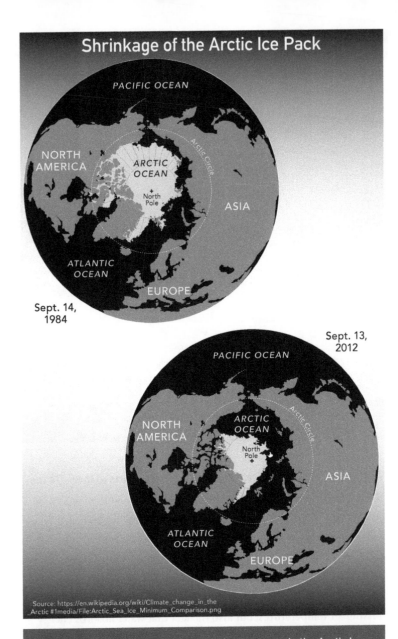

Shrinkage of the Arctic Ice Pack

Sept. 14, 1984

Sept. 13, 2012

Source: https://en.wikipedia.org/wiki/Climate_change_in_the_Arctic #1media/File:Arctic_Sea_Ice_Minimum_Comparison.png

Over the course of nearly thirty years, ice coverage in the Arctic has shrunk considerably. In 1984, the ice coverage was about average for the region up to that point. In 2012, the ice coverage was about half that amount.

combined absorbing only about 4 percent. So, the water is getting hotter, and the rate at which that's happening is increasing too.

Ice that floats on water—such as some Arctic glaciers—heats up when the water below it does, and it melts. With less ice coverage to reflect the sun's energy, that water absorbs even more heat, which can melt more ice. With both warmer air and warmer water acting on them at the same time, many polar glaciers are essentially melting from both sides, squeezed above and below.

How much are they melting? In a landmark study released in 2019, the IPCC found that the Greenland ice sheet lost an average of 275 gigatons (that is, 275 billion tons [249 billion t]) per year between 2006 and 2015. In just one day in July of 2019, the island of Greenland lost two billion tons (1.8 billion t) of ice. That's more than 4 trillion pounds (1.8 trillion kg). At this rate, scientists predict some areas near the Arctic could be completely free of ice during the summer months before the end of this century.

Things aren't great on the other side of the planet either. Per the journal *Nature*, Antarctica lost more than three trillion tons (2.7 trillion t) of ice between 1992 and 2017. The world's largest ice shelf (a slab of ice extending over the ocean) is the Ross Ice Shelf in western Antarctica, and it's melting ten times faster than scientists originally predicted because of the heat from warmer-than-normal ocean water.

Another problem is parts of glaciers breaking off from land and becoming sea ice or icebergs. This process is called calving, and it provides some of the most striking pictures of how much climate change is altering the planet. The pieces of ice that break off can be massive. One the size of Tampa, Florida, broke off from the Pine Island Glacier in Antarctica in February 2020. The temperature at the time was 65°F (18.3°C), which is extremely warm for Antarctica. Pine Island in general is shrinking quickly. It retreated more than 30 feet (9.1 m) per day during the summer of 2019–2020. As glaciers melt, they get thinner, which makes them less stable and more likely to calve. Calving is another reason why glaciers are shrinking by larger amounts than winter conditions can replenish.

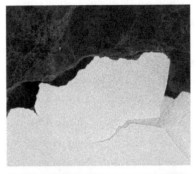

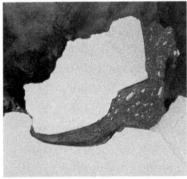

These two photos of the Amery ice shelf in Antarctica show a piece of ice about the size of the greater London area breaking off from the shelf. Scientists who monitor ice calving in Antarctica say this event, which occurred in 2019, was the biggest calving they'd observed in decades.

For nearly a decade, researchers have also noticed that the oldest sea ice in the Arctic is disappearing. There are a few reasons why that's troubling. That this old ice previously managed to last through centuries of summers and winters, only to melt now, shows that current warming isn't normal or part of a natural cycle. The older perennial ice is also thicker because its winter layer forms on top of existing ice, as opposed to the thinner annual ice that forms in winter and melts in summer. Since 1980, according to NASA, the amount of perennial ice has shrunk more than 12 percent per decade. On top of that, the summer melt is also happening earlier and earlier, causing a greening effect in which areas usually covered by ice and snow are now exposed to the sun.

Melting sea ice and icebergs don't *directly* cause the sea level to rise because that ice is already in the water and it fills roughly the same volume in both solid and liquid form. On the other hand, melting land ice like that covering Antarctica adds to the sea level. Whether it melts or calves into new icebergs, it adds more water to the oceans. That makes the oceans rise.

There's a simple experiment you can do at home that loosely illustrates this. Take a glass or aluminum pan, include some "land" in the form of a solid substance that won't melt (such as rocks or wood), and fill the rest of it nearly to the top with water. Put a few ice cubes on top of that land and watch what happens. If it's cold enough that the ice doesn't melt, the cubes will sit there without much change to the water level. As the ice melts, the water level in the container will go up and eventually overflow the sides. On the other hand, if you fill the same pan with only water and floating ice cubes (no "land"), you won't see the same kind of rise when the ice melts.

When that land is, say, Antarctica, the amount of ice involved means that literal drops in the ocean can add up to slightly higher seas. And as we discussed at the start of this chapter, even slightly higher seas have real consequences.

Worst-case scenarios for rising seas are basically doomsday outcomes out of a disaster film (and quite a few movies with that premise have already hit theaters). Losing just the Greenland ice sheet could raise average worldwide sea levels about 20 feet (6 m). That's obviously a dramatic amount. It's higher than you and two friends standing on each other's shoulders, unless you're all centers in the NBA. The Greenland ice sheet also wouldn't disappear on its own; other glaciers would also shrink dramatically. Losing all glaciers worldwide would raise the seas more than 200 feet (61 m). As scientist Neil DeGrasse Tyson has put it, the total melting of all polar ice would cause flood levels that reach as high as the Statue of Liberty's elbow. That's still many years and many bad choices away, but people keep taking steps down that path, and all the stops along the way feature higher seas. The question is how many of those steps we take.

SOME REALLY BAD TIMING

Venice, Italy, is famous for its low elevation. For centuries, its location as a cluster of islands in the delta of two major rivers has been a big part of the city's appeal. Its iconic canals made it an ideal trade hub historically and a popular tourist destination in modern times. More than thirty-five million people visit Venice in a typical year, many of them taking gondola rides through the canals. In November 2019, the city faced its highest water levels in fifty years (and its second-highest in nearly a century). About 85 percent of Venice flooded. The rising water did hundreds of millions of dollars' worth of damage, including to some of the city's most famous buildings.

The Venice city council chamber flooded for the first time in its history just two minutes after the council rejected a series of measures to fight climate change. The sight of the empty chamber full of green water with recently debated papers floating in it became an iconic image of inaction. To be clear, Venice isn't run by climate deniers. Since 2003, the city has been building gates and other infrastructure to try to protect it from high tides, but the city's mayor described the 2019 flood as a wake-up call that Venice—and Italy in general—needs to take greater action to fight climate change.

St. Mark's Square in Venice floods so often that workers build temporary sidewalks to cross the plaza.

Other Rising Concerns

Melting ice is obviously a big deal when it comes to the sea level, but the extra water alone is just the (ahem) tip of the iceberg.

Remember from chapter 1 that polar ice is responsible for reflecting much of the sun's energy back into the atmosphere, instead of absorbing it the way land or water does. When there's less ice, there's less surface area to reflect sunlight and more of the sun's heat will stick around, causing warmer average temperatures around the poles. Warmer average temperatures mean more melting ice and warmer waters, which also melt more ice. With more ice melting, sea-level rise speeds up, and more of the planet is going to flood. And another vicious cycle ensues.

Even in places where there isn't any sea ice, warmer oceans add to the problem of rising seas. When water is heated, it expands and gets less dense. If it's heated from room temperature to its boiling point, water will expand by about 4 percent. That might seem like a small number, but heated oceans don't need to expand by very much to cover low-lying land (especially at high tide), not when that expansion is combined with the other factors raising the sea level. As the oceans keep getting hotter, they expand more, causing sea levels to rise and flooding to increase. As much as half of the sea-level rise of the past twenty-five years was caused by this thermal expansion, and that trend is going to continue.

There are other factors at play too. In chapter 2, we talked about Pacific Ocean weather patterns like El Niño and La Niña (see pages 54–55), which can also alter the sea level. El Niño brings additional warming, so oceans expand during El Niño years. In places already vulnerable to rising seas, an El Niño event can be the little extra push needed to cause more devastating flooding. La Niña does the opposite, causing oceans to contract temporarily. If you look at charts of sea levels over time, they generally go up, but with a few dips right after El Niño years or during La Niña ones. That doesn't mean the water isn't rising overall, just that these weather patterns still have some measurable impacts.

Extra water in the ocean can be destructive in even more ways we haven't talked about yet. Consider tsunamis, which are series of particularly large waves caused by earthquakes, erupting volcanoes, or underwater landslides. The motion from these phenomena gives ocean water a strong push in one direction. Rising seas will make tsunamis a lot larger and more powerful by giving them more sea water to use as fuel.

Two of the most destructive tsunamis in history hit in the past twenty years. The Indian Ocean tsunami that hit on Boxing Day (December 26) in 2004 killed more than a quarter of a million people, most in southern Asia, and featured waves more than 50 feet (15 m) high. A 2011 tsunami killed more than twenty thousand people in Japan. It also damaged several nuclear reactors located near the coast, releasing dangerous radiation into the water and atmosphere. Rising seas will keep making tsunamis more destructive, along with hurricanes and other storms we talked about earlier—not by causing them directly, but by giving them the additional power they need to last longer, dump more water, and kill more people.

It isn't just the ice melting at the poles that contributes to rising seas; smaller glaciers all around the world are doing the same thing. Glaciers can be found (at least for now) on every continent—if you count all of Oceania as a continent rather than just Australia—and on several other land masses. The glacier atop Mount Kilimanjaro in Tanzania is only about one-fifth the size it was in 1912. From Glacier National Park (see the sidebar on pages 96–97) in the United States to the Fox and Franz Joseph glaciers in southern New Zealand to Mount Blanc in the Swiss Alps, glaciers are noticeably smaller than they were not that long ago. Melting glaciers are a worldwide problem.

Iceland, which has been known for its glaciers since Viking times, said goodbye to one iconic example in 2014, when the shrinking

Okjökull passed the point where it could still be considered a glacier. Scientists there held a "funeral" for Okjökull a few years later, including the placement of a memorial plaque warning that all Iceland's glaciers were on track to be gone by 2200 and telling future visitors, "This monument is to acknowledge that we know what is happening and what needs to be done. Only you know if we did it."

Add all of them together, and glaciers worldwide are already losing more than 350 billion tons (317 billion t) of ice and snow each year, at a rate about five times as fast as they were melting in the 1960s. And a lot of that melting water eventually makes its way to the oceans, adding to the overall issue of sea-level rise.

Water from melting inland glaciers can also lead to more precipitation in the surrounding areas, which floods rivers, lakes, and other nearby bodies of water. And as the glaciers disappear, they stop freezing in winter and melting in summer, taking away an important natural resource for people who rely on that summer melt for water to drink and to irrigate their crops. For example, summer melt from the Himalaya mountains has, for centuries, formed many of the rivers that provide water to places in China and India, where many millions of people rely on them. In places where drought is already an issue, losing that water source wouldn't help.

Of course, as glaciers cover less of the planet's land, there's even less surface ice to reflect the sun's energy and prevent more warming. The effect of this on Earth's permafrost could become the most dangerous aspect of melting. As the name suggests, permafrost is a thick layer of soil beneath the Earth's surface that stays permanently frozen. It exists in places where the temperature rarely rises above freezing, such as in far-northern parts of Alaska, Russia, and Greenland. It covers nearly 9 million square miles (23 million sq. km) and can be as thick as 3,200 feet (975 m) in places. And it's thawing too. This thaw started slowly a few decades ago and really sped up in the past few years. In 2018, researchers in Siberia found permafrost thawing even during the winter, which was obviously concerning (few things are as notoriously cold as Siberian winters).

One of the big problems with melting permafrost is methane, the greenhouse gas that is about twenty to thirty times more powerful than carbon dioxide. A lot of methane is trapped underground in the permafrost and some of it has been frozen there for thousands of years. Melting permafrost can let that methane escape into the atmosphere—and too much of that could be the last straw for life on Earth. Even now, big methane bubbles are visible under thin ice in places where permafrost has melted, and some of the gas has already leaked out and entered the atmosphere. There's a lot of carbon dioxide frozen in the permafrost too. Many future climate projections, including some of the IPCC ones from earlier in this chapter, didn't take melting permafrost into account because it wasn't yet clear that it would happen.

The entirety of Earth's permafrost isn't at risk of disappearing yet, but even a small amount of it melting away is a big problem. There are various estimates about how much the global temperature would need to rise before enough of the permafrost would melt to allow the methane underneath to escape in large quantities. But it's a good idea to stop this from happening before we find out which estimate is right.

Rising All Over

Like its land, Earth's oceans aren't all impacted the same way by climate change. Areas closer to the equator experience higher temperatures, and expand more, so sea-level rise is usually worse in those areas than in places closer to the poles. Sea levels can also be different by area depending on local factors such as wind patterns, ocean currents, and the depth of the ocean.

Still, rising waters will affect a wide range of places. The low-lying island countries discussed at the start of this chapter are probably going to be the first ones made uninhabitable by rising waters, but this is partly because they're so small. Other, bigger countries could lose more total territory (and possibly more people) to the rising seas. They just have enough room elsewhere that they aren't at risk of disappearing.

If the sea rises as much as currently predicted, the combined land it will permanently flood around the world by the end of the century is home to about two hundred million people. That's more than 2.5 percent of the planet's population. Another one hundred million people live in places that will have to deal with chronic, regular flooding.

A 2018 study by the World Bank looked at the risks for the Asia-Pacific region and found China and Indonesia stand to lose the most land area to permanent water. The study estimated how much land would be lost if the sea level rose either 3.3 feet (1 m) or 9.8 feet (3 m) by the end of the century. For China, the first scenario would see 12,355 square miles (32,000 sq. km) along the coast at risk by 2100, an area where more than twenty-three million people live. If we hit the second scenario, that goes up to 27,400 square miles (71,000 sq. km), affecting fifty-two million people. That's more people than live in any single US state. It's about the size of California with enough room left over for Ohio. For Indonesia, 3.3 feet of rise puts 5,300 square miles (13,800 sq. km) at risk, and 9.8 feet (3 m) would impact 13,500 square miles (35,090 sq. km).

With more than 168 million people as of January 2020, Bangladesh is the eighth-largest country in the world by population. Those many millions of people are packed into an area a little less than half the size of the United Kingdom. Bangladesh is also one of the least-elevated places on Earth because much of its land is part of a large river delta. Two-thirds of the country sits less than 16.4 feet (5 m) above sea level, and one-quarter is barely 6.6 feet (2 m) above it. It's considered one of the most likely countries to lose much of its land to rising seas and having that many people potentially affected makes Bangladesh an obvious source of future climate refugees (more on that in chapter 6). In the meantime, rising water levels—sometimes combined with extreme weather like cyclones and stronger-than-normal monsoons—are already battering the Bangladeshi coast. In 2019 alone, more than three hundred thousand people there were displaced from their homes during July floods.

Some of the biggest cities in the world also happen to sit along coasts that are vulnerable to rising seas. Waterways were always natural places to build cities, giving people direct access to trade and travel routes, along with beautiful views. The historical advantage of coastal locations is another thing the climate crisis is erasing.

More Americans live in New York City than in any other metropolis (the runner-up, Los Angeles, has fewer than half as many people). More than one-third of lower Manhattan could be at risk of submersion by 2050 if things continue as they are moving. City officials have been studying adaptation plans for more than a decade, and in 2019 the mayor announced plans for a multibillion-dollar project to try expanding the island's coastline and building physical barriers as ways to protect the city.

THE WATER WHERE YOU LIVE

We've talked about a lot of places on Earth that are or will be affected by rising seas. But what about your hometown?

Whether you live in the United States or not, NOAA has a useful tool to check out. Its Sea Level Rise Viewer (https://coast.noaa.gov /slr/#) lets you adjust variables like the amount of sea-level rise or the year and to view local scenarios by zooming in anywhere on a map. It probably won't paint a rosy picture, but it's a good way to visualize how the issues covered in this chapter will alter places you know.

Mumbai, India, is home to more than eighteen million people—a larger population than that of more than 130 countries. Along with the annual monsoons we discussed earlier, this coastal city is also being flooded by rising sea levels, and much of it will be underwater by the end of the century if things don't change. Many other cities near coasts face the same poor odds: Miami, Bangkok, Tokyo, Shanghai, Kolkata, and New Orleans to name a few. As with island nations, it'll become difficult, and eventually impossible, for people to live in those cities even before they are covered by ocean water.

Coastal cities and towns are already seeing more flooding. NOAA tracks sea-level rise around the United States using nearly one hundred tidal gauges and other tools to measure changes in the water's level. It found that 2015 was a record-setting year, only for 2018 to match the record, with five days of high-tide flooding nationally, and specific regions seeing even more.

We've now discussed how climate change is altering Earth's physical environment. In the next chapters, we'll talk even more about what those changes mean for life on the planet—including people.

CHAPTER 5

LIFE DURING WARMING TIME

We are supporting and subsidizing the very things
that are damaging our planet. The natural world
is so delicate. It needs all the protection it can get.
Sometimes that means governments have to take
decisions that are painful and cost money.

—Sir David Attenborough, acclaimed naturalist and broadcaster,
in a 2019 speech to the International Monetary Fund

EUROPEAN SAILORS FIRST spotted the Bramble Cay
melomys, also known as the mosaic-tailed rat, in the 1840s. A
mouse-like animal with a longer tail and bigger feet than many of its
relatives, the melomys lived on a small island near the northern tip of
Australia's Great Barrier Reef and not far from Papua New Guinea.
There, it didn't face too much competition from the green sea turtles
and various seabirds that lived there or stopped by to lay the eggs that

were part of the melomys's diet. As recently as the 1970s, Bramble Cay was home to hundreds of these little animals, and later surveys suggested the population was still doing fine.

That's sadly not true anymore. Sea-level rise destroyed nearly all of the melomys's habitat, as the tiny island where they lived is less than 10 feet (3 m) above sea level at its highest point, and the ocean in that part of the world has risen about twice as fast as the oceans overall. Just between 2004 and 2014, floods and storms wiped out 97 percent of the island's vegetation, washing away the plants the rodents needed for food and shelter and probably a number of the animals themselves. After nearly a decade without a single sighting by scientists, the Bramble Cay melomys has the unfortunate distinction of being the first mammal species officially listed as extinct because of human-made climate change. Researchers classified it that way in 2015, and the Australian government followed suit four years later.

Because it wasn't a particularly famous animal, you might not have known about its disappearance. But to scientists who had been warning about the risk rising sea levels posed to this particular melomys species, only to see it go extinct pretty quickly, it was further confirmation that climate change is already altering life on Earth in permanent ways. The melomys may have been the first mammal lost to human-made climate change, but it probably won't be the last, and other kinds of life are facing similar challenges. Even those plants and animals not at risk of disappearing are going to have to adapt, which will impact other species—including our own.

Fighting for Survival

We're living through what scientists consider the planet's sixth mass extinction since life on Earth began (for more on the other five, see sidebar on pages 112–113). This ongoing mass extinction dates back to the end of the Ice Age less than twelve thousand years ago. It goes back to the mammoth and mastodon, through the dodo and passenger pigeon and Tasmanian tiger and on to the present. Basically, it's the era of humankind causing the extinctions of other species.

Human influences such as habitat destruction, hunting, competition for resources, and introducing invasive species have consistently been major parts of the problem. But like hurricanes or

EXTINCTION DISTINCTIONS

What life on Earth is going through now is often called the sixth mass extinction or the Holocene extinction. Here's a rundown of the first five.

First: Ordovician-Silurian extinction
444 million years ago
This extinction was caused by a different kind of climate change than what we are experiencing. The globe cooled dramatically, and sea levels fell hundreds of feet as glaciers spread, killing tons of sea life. Nearly all of the species on the planet at the time (early versions of coral, algae, and fish, plus invertebrates like trilobites and brachiopods) lived in the oceans, and about 85 to 90 percent of all species disappeared.

Second: Late Devonian extinction
About 383 to 359 million years ago
This extinction took place gradually, over more than twenty million years, so it's hard to blame it on a specific cause. Any or all of the following may have played a role: a volcanic eruption pumping greenhouse gases into the atmosphere, dropping oxygen levels, an asteroid, and changing sea levels. About three of every four species around at the time went extinct.

Third: Permian-Triassic extinction
About 252 million years ago
Serious global warming was at fault here, sparked by the eruption of a volcanic region in what's now Siberia. The eruption dumped

droughts, this situation has been getting a lot worse as climate change alters the world, and the rate of extinction has been speeding up dramatically in the past few decades.

In May 2019, the Intergovernmental Science-Policy Platform on Biodiversity and Ecosystem Services (IPBES), released a report saying that at least one million species—about one of every eight on Earth—are already threatened with extinction because of human activity. Let's put that number in perspective. Scientists estimate

more than 14 trillion tons (12.7 trillion t) of carbon into the atmosphere, more than two and a half times as much carbon as all of Earth's fossil fuels could produce. This "great dying" was the most destructive event in the planet's history and wiped out almost all sea life, the planet's forests, and three of four land species in the span of about sixty thousand years.

Fourth: Triassic-Jurassic extinction
201 million years ago
Another case of climate change, brought about by more volcanic activity, was probably the cause here. Earth grew between 5°F (2.8°C) and 11°F (6.1°C) hotter, and the greenhouse gases made oceans acidic. About 80 percent of life on Earth at the time disappeared, including many ancient crocodilians and sea reptiles. This extinction created the conditions in which dinosaurs would soon become the planet's dominant lifeforms.

Fifth: Cretaceous-Paleogene extinction
66 million years ago
Odds are you know about this one—this is the end of the dinosaurs. The likely culprit was an asteroid that hit Mexico's Yucatan Peninsula, forming a crater about 120 miles (193 km) wide. The dust and debris caused by the impact blocked the sunlight, cooling the planet and killing plant life, leading to a domino effect that eliminated three of every four species, but giving smaller lifeforms including birds and mammals an opening to take over.

that there may be about 8.7 million total species of plants, animals, fungi, bacteria, and other life on Earth, but we've only identified, categorized, and named about 1.6 million species. (Scientists identify more all the time; in 2019 alone, they identified seventy-one previously uncategorized species.) Every living thing you or anyone else can identify is contained in that 1.6 million, and the IPBES study said nearly that many species are already at risk. To understand the scale, think of every species you can name possibly disappearing in your lifetime, and know that still understates the crisis.

As the IPBES put it, the global extinction rate by 2019 was already "tens to hundreds of times higher than it has averaged over the past 10 million years." It based that conclusion on the results of more than fifteen thousand scientific and governmental studies around the world. Representatives of 109 nations approved the report, which made it one of the most thorough on the subject. Its conclusion? "Human actions threaten more species with global extinction now than ever before."

Climate change isn't the sole driver of all those cases, but it's a primary or secondary factor for many of them. And in other cases, environmental issues that play a role in climate change—from causes such as deforestation and pollution to effects such as extreme weather and drought—are at least partly responsible for decreasing species' chances of survival. Life on the planet has always needed to adapt to Earth's climate, but the climate is now changing too fast for many species to do so. About half of all animals on Earth have already seen their habitats, food supplies, and populations impacted by climate change. Cases like that of the Bramble Cay melomys, in which the effects of human-driven climate change become irreversible, are going to become more common if things don't improve. And every extinction has a domino effect. All species have a role in their ecosystems, and other species need to adapt any time one goes extinct. If they can't, they vanish too.

Some endangered species, including several of the world's most well-known animals, face risks from these domino effects of climate change. Tigers, giant pandas, orangutans, Asian and African elephants, mountain gorillas, and cheetahs, among many others, are already listed as vulnerable or endangered because of their limited available habitat and other interference from people. We obviously can't cover the situations of more than a million different creatures in this book, but let's briefly talk about a few of them.

Animals Crossing into Danger

The Bengal tiger is the second-largest tiger subspecies. It is the national animal of both India and Bangladesh, two countries facing serious struggles with climate-change impacts (see pages 52 and 107, respectively). The marshy mangrove forests where many of the remaining tigers live, known as the Sundarbans, are shrinking because of rising seas, and the area is vulnerable to cyclones and extreme weather. About 70 percent of that important land is only slightly above sea level, so tigers will need to find a new habitat as it floods. Habitat destruction also affects the animals that tigers prey upon, such as sambar deer, wild hogs, and hog deer, causing tigers to look for other food—and that can include farm animals. In developing countries with large populations and fast-growing cities, tigers will struggle to find the new habitat they need without competing with people more than they already do. Nobody wins that competition.

Because there are only a few thousand tigers—of all subspecies, including the Bengal—left worldwide, they've become a focus of conservation efforts, and conservationists have taken steps to protect their habitat. Illegal hunting definitely doesn't help, nor does the illegal wildlife trade. Habitat destruction is such a threat that the majority of the world's tigers live in captivity, and additional habitat loss could eliminate the remaining wild populations of tigers.

Amphibians are another set of animals at risk. About one in three known amphibians, about eighteen hundred species, already faces the threat of extinction. And climate change is hitting them in a couple of ways.

Warm-blooded animals like birds and mammals regulate their own body temperatures, but ectotherms, or cold-blooded animals, get their body heat from their environment. That's all well and good when temperatures are within the normal range to which the animals have already adapted. As the climate gets warmer and hot days get hotter, though, animals can literally overheat. That's especially true of amphibians like frogs, toads, and salamanders because of their skin. Amphibian skin is unusually thin and permeable. Water, oxygen, and heat can get through it easily; it literally helps them breathe. But they also need moist environments to survive. Already, amphibians around the world are at risk because of habitat destruction, and the worsening droughts due to climate change are reducing that habitat even more. The pollution that's damaging the atmosphere is also harming amphibians because the animals can directly absorb air and water pollution through their skin.

The lemur leaf frog (*Agalychnis lemur*), a tree frog native to Central America's rain forests, is at risk of extinction due to a fungal disease that is made worse by the changing climate.

If that wasn't bad enough, research suggests that warmer temperatures are helping spread certain fungi species that are especially deadly for amphibians. Two of those fungi,

Batrachochytrium dendrobatidis and *Batrachochytrium salamandrivorans*, attack parts of the animals' skin, making it hard for them to breathe or control the amount of water in their bodies. The fungi eventually cause the amphibians to have heart attacks. On top of that, the fungi spread easily, both through contact between amphibians and through the water where they spend a lot of their time. Even a decade ago, scientists were concerned about the crashing numbers of frog species, including those lost to this fungi pandemic. Some ninety species of amphibians have already gone extinct because of the fungi, and more than one hundred others have seen their numbers reduced by more than 90 percent, making extinction a distinct possibility.

The extreme 2019 Australian bushfires we talked about in chapter 3 are another perfect example of how climate impacts can push species toward extinction. They wiped out huge numbers of individual plants and animals, including many that were already endangered. The fires burned at least 80 percent of the habitat of forty-seven different plant species, including rare ones such as the leek orchid and nightcap oak. Some species of skinks, cockatoos, and turtles lost at least half of their habitat in the fire, and we don't yet know if enough of the animals survived to sustain their populations.

The Kangaroo Island dunnart, a small carnivorous marsupial, was already critically endangered before those 2019 bushfires. Only about five hundred remain in the wild and are confined to just one island, a protected nature reserve off the southern coast of Australia. The fires burned nearly its entire habitat and kept burning for a while. At least a few dunnarts survived in three spots on the island, and more might yet be found, so it isn't extinct—but the species' long-term prognosis isn't good. Another Kangaroo Island native, the little pygmy possum, was briefly feared extinct until a survivor was found in December 2020. More than a dozen species of birds living on the island also saw most of their habitat wiped out, a similarly discouraging situation.

Bushfires don't get that extreme and deadly without the droughts climate change helped create.

In 2018, eight bird species joined the list of animals extinct in the wild. The best known is probably the Spix macaw (the kind featured in the 2011 animated film *Rio*). The bright blue parrot fell victim not just to the illegal wildlife trade but to habitat loss, thanks in part to the same deforestation and wildfires that are

DRILLING DOWN AND MESSING UP

Sometimes one human activity can both cause additional climate change and endanger other species at the same time. For a good example of that, look no farther than Alaska and the push to drill for oil in the Arctic National Wildlife Refuge (ANWR).

The federal government set aside the land in northeastern Alaska in 1960 and later expanded it, protecting an area of more than 30,000 square miles (77,700 sq. km), about the size of the Czech Republic. It is indeed a refuge for a lot of wildlife: bears (polar, black, and grizzly), caribou, musk oxen, moose, wolves, foxes, weasels, beavers, and more. It includes different types of terrain, from evergreen forest to tundra to lagoons and marshes along the coast.

The ANWR also sits on top of large oil reserves. How much oil isn't known for sure, but estimates average around ten billion barrels—an amount the country typically uses in less than two years. Since the ANWR's creation, oil companies have lobbied Congress to open the land to drilling, but they repeatedly failed in those efforts until 2017 when Congress included ANWR drilling in a seemingly unrelated tax bill and Donald Trump signed it into law.

If the oil is extracted and burned, it is obviously going to contribute more to climate change. Losing pristine wilderness forever in exchange for less than two years' worth of oil supply is a shortsighted trade (and the ANWR has already experienced melting ice and other climate-change effects). The oil drilling itself also threatens the area's wildlife. Energy companies use infrared technology to find (and avoid) polar bear dens, which sounds great, except that studies show it fails to locate those dens more than half the time, which means already endangered cubs can be killed in their shrinking the Amazon rain forest in Brazil (see page 83). The ongoing destruction of the Amazon was also responsible for the extinction of other birds, including the cryptic treehunter and Pernambuco pygmy-owl.

The International Union for Conservation of Nature has kept an updated Red List of Threatened Species (that is, species at risk of going extinct in the foreseeable future) since 1964 and cited more

dens or forced to leave them before they're ready to survive in the wild. The coastal drilling will also impact the Porcupine caribou herd, taking over part of the already-small area where the reindeer give birth and raise their young. Thanks to the ANWR's protection from most human activity, that herd is one of the few in North America that hasn't already lost massive numbers.

How many animals and species will suffer from ANWR drilling will depend on exactly how and where that drilling happens, but even the best-case outcome is a huge risk for very little potential reward.

A view of the Arctic National Wildlife Refuge. a huge area of protected land that spans part of northern Alaska. The refuge's future has been a subject of political debate for years.

than thirty-one thousand already at risk. You can check out the ever-changing list and read up on the individual species included at https://www.IUCNRedList.org.

Spoiled Waters

In the last chapter we talked a bit about how oceans are getting warmer and how they absorb greenhouse gases. They're also getting much more acidic because of all the carbon they're absorbing.

Some of the carbon dioxide that dissolves in and interacts with water creates carbonic acid. This process has changed the pH level, or level of acidity, of the ocean. The pH scale goes from zero (pure acid) to fourteen (pure base), with anything below seven more acid than not. Prior to the Industrial Revolution, the oceans had a pH of around 8.18. Since then, it has decreased to 8.07, with most of that change having occurred since the 1990s. As our oceans' pH of 8.07 moves toward 7, they get closer to fitting the definition of *acid*. Also, while these numerical changes seem small (after all, the difference between 8.18 and 8.07 is only 0.11), the pH scale is logarithmic, which means tiny changes in pH reflect much larger real-world changes.

Just as life on land adapted to gradual climate changes but is struggling to adapt to the rapid changes we see now, ocean life is fighting to adapt to warmer and more acidic waters. Those dual problems are proving deadly to a wide variety of marine species.

Coral reefs are one of the clearest examples of this. Corals are animals, living invertebrates related to sea anemones and jellyfish. They live in colonies and many secrete skeletons made of calcium carbonate, which forms a hard, bony structure. Coral reefs are made up of many coral colonies, formed by thousands of coral polyps and the algae they consume.

Reefs are popular with snorkelers and scuba divers because of all the colorful varieties of fish and other sea life (anemones, starfish, and more) that live in them or rely on them for food. Around a quarter of ocean species spend at least some time in coral reefs, even though reefs cover just a fraction of 1 percent of the planet's surface. They're ecosystems no less diverse than rain forests. Corals have survived on Earth for hundreds of millions of years by evolving some unique survival mechanisms, including the ability to rebuild their skeletons and to form reefs. But that natural process can't keep up with the rate of warming.

Because of where they grow, coral reefs are also an important defense for coastal areas, protecting them from tides, storms, and some of the other climate effects we've covered. The Great Barrier Reef off the northeastern coast of Australia is often described as the largest living structure on Earth and is the living thing astronauts can see most easily from outer space. Despite the name, it's actually a system of more than twenty-nine hundred connected reefs, including passages and hundreds of small islands called cays, that is about 1,400 miles (2,250 km) long. The Great Barrier Reef is the biggest reef system on the planet, but there are huge coral reefs in Indonesia, in the Gulf of Mexico, in the Red Sea, and in lots of other places. Climate change is already killing these reefs.

One big problem is coral bleaching. Coral normally has a symbiotic (mutually beneficial) relationship with certain algae species that live within their bodies. The algae give the coral nutrients, as well as their characteristic colors. When the water is too hot, however, the algae living in the coral dies off, and then the coral turns white as if it had been bleached. As recently as the early 1980s, coral bleaching was rare, but a warming planet has brought unusually warm seasons. The first mass bleaching of the Great Barrier Reef didn't happen until the late 1990s; it has experienced five mass-bleaching events since then. A 2016 bleaching killed about 30 percent of the Great Barrier's coral. The reef can still be seen from space, but miles of it are white, and many of the sea organisms that relied on it have needed to find new spots or died off. Bleached corals don't always die, but bleaching, like any serious illness, makes death much more likely. And according to NOAA, by 2050 98 percent of the world's coral reefs will face bleaching each year from higher ocean temperatures unless we cut greenhouse gas emissions dramatically. Some may still survive, but it's a huge risk.

Along with heat, the level of acidity in the ocean can cause coral to deteriorate and also makes it hard for their skeletons to form in the first place. One reef in Hawaii has already hit a tipping point, where the coral's skeletons are dissolving faster than they can repair themselves, shrinking the reef in the process. Scientists have been warning about this for two decades, and the current rates of warming and acidification suggest many more coral around the world will hit that same tipping point before the end of this century.

Just as acidic ocean water partially dissolves coral, it does the same thing to the shells of certain sea animals, including species of clams, conchs and other sea snails, sea urchins, and oysters. They need their shells for protection from predators and other hazards. (Acidic water makes it harder for them to grow shells as well).

The changing climate is also causing big problems for phytoplankton, single-celled organisms that live near the surface of both saltwater and fresh water. Most are too small for humans to see without a microscope, and they come in a variety of shapes and forms. Many of them, such as algae, are plants, but phytoplankton also include protists and bacteria.

Phytoplankton are disappearing because of extreme heat. That's bad for a bunch of reasons. For one thing, they provide food—directly or indirectly—for most of the world's sea life. Just as importantly, about half of the oxygen on the planet—the oxygen we and other animals need to breathe—is produced by photosynthesis in phytoplankton (with trees, grass, and other plants producing the rest). Phytoplankton take in quite a bit of carbon during the photosynthesis process, so we need them to help reduce (or at least stabilize) the amount of greenhouse gases in the atmosphere. Scientists have demonstrated an inverse relationship between the amount of phytoplankton in the water and the amount of carbon in the atmosphere: one goes up, the other goes down. The problem is the global phytoplankton population is only around 60 percent as high as it was in 1950.

Warming oceans have also affected where phytoplankton live. And as parts of the oceans are getting too hot for them to survive, they're moving to colder waters, with their populations shifting away from the equator both north and south, which will also shift the habitats of the sea animals that rely on phytoplankton for food. Zooplankton, which are tiny animals, are also in trouble. Their populations are shifting to more suitable waters as well, and those with shells have the same problem with acidic water that other shelled animals do.

Even krill, one of the most abundant kinds of ocean life, aren't safe from climate impacts because they rely on phytoplankton for much of their diet. Krill are themselves a key part of the food chain, as the little crustaceans serve as a popular meal for penguins, seals, and baleen whales—the ones without teeth, such as blue, right, and humpback whales. A single blue whale can eat a few tons of krill per day. When there's not enough phytoplankton, the consequences ripple through the food chain from the smallest to the largest creatures on Earth.

Disappearing sea ice and shrinking glaciers are a big problem for wildlife too. If there's an animal that's become the symbol of climate change, it's probably the polar bear. The bears are comfortable in icy water, but only for so long. They can swim for hours and for dozens of miles, but they still need solid surfaces (often in the form of sea ice) to rest, eat, mate, and raise their cubs. They also need sea ice to hunt for food such as seals and walruses. As sea ice disappears, some bears starve, while others shift their habitats. One long-term study in northern Canada found the percentage of polar bear sightings on land or water (as opposed to ice) increased from about 12 percent in the 1980s to 90 percent from 1997 to 2005. As bears journey to new

THE TIDE IS . . . RED

While most phytoplankton don't fare well in hot water, there are some exceptions. Specific kinds of algae do great in those conditions, including some extremely toxic ones. And that's bad news for many other animals.

When toxic algae bloom—that is, when they quickly reproduce in big numbers—they can create what's commonly called a red tide, which looks like what it sounds like. The more accurate scientific term for these events is a harmful algal bloom (HAB). (All algae bloom, but the non-toxic variety aren't harmful.)

These kinds of blooms have been reported for centuries, but they seem to be getting worse. Thousands of animals were killed by red tides off the coast of Florida that started in 2017 and lasted more than a year. The governor declared a state of emergency for several coastal counties. Rust-colored tides brought the bodies of fish, dolphins, and sharks to shore, along with already endangered manatees and sea turtles. Toxic algae can also produce other colors, like the blue-green cyanobacteria that covered most of Lake Okeechobee, Florida's biggest freshwater lake, that same year.

Blooms affect people too. A bloom in Lake Erie in 2014 badly poisoned drinking water in parts of Ohio. The toxins that kill fish also make them poisonous to anyone who eats them. Every coastal state in the United States has experienced an HAB, which can also make the air near the coast hard to breathe.

Whether warming oceans are causing red tides or just allowing them to thrive and spread isn't completely clear. HABs seem to get worse after extreme storms, and fertilizer and other nutrient-rich runoff from flooding have played a part in some HAB events. One thing is clear, though: they're another serious pressure on a struggling ocean environment.

Red tides caused by harmful algae blooms cause waters to turn red. The algae kill huge numbers of sea animals. from fish to turtles to dolphins.

places in search of food, they've encroached on towns in Russia and Canada more often. Polar bear populations are generally estimated to be higher than they were in the 1960s because of several conservation laws enacted to save them (most notably an international 1973 agreement to restrict hunting). But melting sea ice threatens to undo that progress. If current trends continue, their population could drop by about two-thirds by 2050.

It's not just endangered animals that suffer. The 2019 documentary series *Our Planet* made international news with its footage of walruses in northern Russia overcrowding on top of steep cliffs. Those who couldn't find room plummeted down the mountain to their deaths, landing near others who met a similar fate. Those brutal deaths have become more common as retreating sea ice leaves the huge marine animals without enough territory. Walruses falling off cliffs has been documented before, but it was previously because something disturbed the group, such as a polar bear or a helicopter, causing them to panic and fall, rather than the animals not being able to find enough room.

Though walruses spend a lot of time in the water, they need surfaces to rest on just like polar bears do. Walruses spend most of their time on sea ice. Many thousands of them can join together in large groups known as haul outs. When sea ice isn't available or coastlines are smaller than they used to be because of erosion and sea-level rise, those groups resort to cliff sides and other limited spaces. In addition, females with new calves usually spend Arctic summers on sea ice, far away from the shore. As summer ice becomes harder to find, they're traveling to the shore and joining large groups that don't have space for them. Although the species as a whole isn't yet at risk, individual animals still suffer.

Changing Places

Unlike humans, other animal species haven't spent the past few decades bickering about whether the climate was changing or whether that was worth doing something about. They reacted to it. They had to. Same with other forms of life on Earth, from plants and fungi to bacteria and other microscopic organisms. For a lot of them, that's meant moving somewhere they can survive.

As many as half of the planet's species are already on the move. In 2017, the lead scientist of a *Nature* study put it bluntly: "We're talking about a redistribution of the entire planet's species."

In Great Britain alone, more than fifty land-based animals have already shifted their habitats just in the past decade. That includes birds such as purple herons and European bee-eaters, the purple emperor butterfly, and many species of flies and bats.

The American pika, the smallest of the lagomorph family, which includes rabbits and hares, has survived in extremely cold, Arctic, mountainous environments for ages. But as those environments get hotter, pikas have begun to completely disappear from some parts of their old range and shift to living at higher elevations in the mountains where it's still cold enough to get by. Scientists consider these pikas one of the mammals most likely to disappear because there's a limit to how much higher they can move as their habitat warms.

Monarch butterflies, one of the most recognizable of butterfly species, have seen their numbers plummet in a short amount of time in parts of their old habitat. One area in the western United States used to get about 4.5 million of the butterflies during their migration every winter; in 2019, that number was only about thirty thousand. Not only are butterflies sensitive to temperature changes, but so is milkweed, a main food source for the monarchs and their larvae. At certain temperatures, some varieties of milkweed can become poisonous to the butterflies, which has started happening more frequently.

Mackerel in the Atlantic Ocean have always migrated during the year, but their migration pattern has moved farther north as the oceans have warmed. They're now regularly caught in Iceland, which didn't happen just a few years ago. Lobsters along the eastern United

States coast have already moved their range north too. Some species are relocating by hundreds of miles. For countries that rely on fish or seafood supplies for their economies, seeing the animals move into other nations' territorial waters is a challenge even if the species aren't endangered.

Animals aren't alone in doing this; plants are shifting too. By the end of this century, the climate around Paris, France, is expected to be closer to that of Madrid, Spain, and that same shift is putting some iconic French grape varieties at risk. Signature international foods like olives, wine, and coffee—items closely associated with the regions where they grow—are already struggling in some areas that have grown too hot. That's bad news for growers who have cultivated these climate-sensitive foods for centuries (and for the food connoisseurs who love them), and production will have to keep shifting north in the years ahead. It's yet another example of a problem we're already seeing that is probably going to get worse.

Getting Sick from the Heat

The deadliest animal on Earth for humans isn't some massive predator or poisonous snake or even other people. It's the mosquito.

Mosquitos kill about three-quarters of a million people every year—a little more than the population of cities like Denver or Seattle—because of the diseases they carry. Most people who live in a mosquito habitat have been bitten at some point and know that the insects feed on blood. They also ingest viruses, bacteria, and other parasites from the animals they bite and then spread those when they bite other animals (including people).

Malaria is one of the most widespread deadly illnesses in the world. It infects more than two million people per year and kills anywhere from four hundred thousand to six hundred thousand. That's where most of mosquitos' human death toll comes from, but mosquitos can also carry yellow fever, dengue fever, West Nile virus,

Zika virus, several forms of encephalitis, and other diseases you definitely don't want to get. And it isn't just humans that get sick from mosquito-borne illnesses—other mammals, birds, and even amphibians can die from them too.

While other species are suffering because of warmer temperatures, mosquitos are doing great. They thrive in warm and humid weather, which is why they're more common in summer and closer to the equator (there's a reason why the area called the Mosquito Coast is in Central America). As the world heats up, mosquitos can get to places they never or rarely could before—both in terms of their range and the elevation at which they can survive—and do so for bigger chunks of the year, putting more people and animals at risk.

The Hawaiian island of Kaua'i is home to several species of honeycreepers—brightly colored birds that used to be widespread in Hawaii for millions of years. Most species have since gone extinct, with habitat destruction one factor. But another cause was mosquito-borne diseases such as malaria and the avian pox. Mosquitos did not make it to Hawaii until well into the 1800s, so most honeycreeper species weren't prepared for them and the illnesses they carried. As of 2009, according to the US Geological Survey, seventeen species of honeycreepers were already extinct and fourteen others were endangered. Those still surviving live in high altitudes in Hawaii's mountains, where it was cold enough that mosquitos weren't an existential problem. That is, until warmer temperatures put 60 to 96 percent of that elevated area at risk for mosquito invasion, allowing mosquitos to move to higher elevations and do so for longer periods.

The long-term survival of honeycreepers is going to come down to whether climate change can be slowed and whether people can actively keep mosquito larvae from growing at higher altitudes.

Ticks are spreading too, as warmer temperatures let them survive in places they couldn't before. This poses a problem in the United States and Canada, where the tiny arachnids carry the bacteria

FOR THE BIRDS

The non-profit Audubon Society has been studying birds and working to protect the birds of North America and elsewhere for more than a century. And it's been issuing some serious warnings about what climate change is doing to many bird species. In its 2019 study "Survival by Degrees," the organization identified 389 North American species at risk due to climate change—roughly two-thirds of the continent's feathered inhabitants.

Based on that information, Audubon launched its Birds and Climate Visualizer, which lets users in the United States, Canada, and Mexico see which birds are at risk in their own state or province—or even their county or zip code—and how vulnerable different birds are. It also shows how the numbers of at-risk birds change based on how hot the planet gets, based on increases of 2.7°F (1.5°C), 3.6°F (2°C), and 5.4°F (3°C). At each of those levels, some bird species will need to relocate, and some will not survive, but keeping the temperature as low as possible is the best bet to save them.

You can try the Birds and Climate Visualizer yourself at: https://www.audubon.org/climate/survivalbydegrees#climate2 -survival-search.

responsible for Lyme disease, which causes rashes, fatigue, and long-term joint pain if it isn't treated quickly enough. Ticks need host animals, such as deer and other mammals as well as birds, to complete their life cycle. Because several deer species are among the animals whose habitat is shifting north as the climate warms, they're bringing the parasites to places where people and animals aren't used to dealing with them and the diseases they bring. And Lyme disease is already becoming common in places it wasn't before.

There are plenty of other examples of pests—those that carry disease, those that destroy crops, and those that do both—finding more warm places to live and do damage if climate change keeps going the way it is. But that's just one way climate change is hurting people. We'll look at some others in the next chapter.

CHAPTER 6

A CHANGING SOCIAL CLIMATE

The consequences for humanity are grave.
Water scarcity threatens economic and social gains
and is a potent fuel for wars and conflict.

—Ban Ki-Moon,
former secretary general of the United Nations, in 2007

DHAKA, THE CAPITAL of Bangladesh, is one of the world's ten biggest and fastest-growing cities. About twenty-one million people live there, and its population more than doubled between 1990 and 2005. To put that in perspective, it's home to about twice as many people per square mile as the island of Manhattan.

Some of that growth has been the kind that cities like to see. Dhaka is the economic and media center of Bangladesh, with a huge textile industry and a growing middle class. On the other hand, another part of its growth comes from the massive slums that have formed around the city as people move there from other parts of the country. As climate change destroys coastal towns and forces Bangladeshis to relocate, many head to the capital in search of work.

More than 70 percent of the people living in Dhaka's slums moved there because of environmental damage somewhere else. That population increase is creating massive congestion, straining city services, and generally flooding the city with more people than it can handle—on the order of four hundred thousand migrants per year, just from elsewhere in the country. Businesses are trying to create more jobs to take advantage of so many available workers who will work for little pay, and other Bangladeshi cities are developing plans to attract climate migrants there instead. Even if they succeed, the number of people losing their homes to climate change means more people on the move and more strain on cities.

As we discussed in chapter 4, Bangladesh is one of the lowest-lying countries in the world, making it particularly vulnerable to rising seas. Annual flooding destroys homes along the coast, and the salt from that flooding damages crops every year. The country also gets hit with annual monsoons like the ones mentioned in chapter 2. Those problems add up and put Bangladesh in a difficult predicament. It's only about the size of Illinois but, with a population of roughly 165 million, has more people than all but seven other nations. If the sea level there rises just 3.3 feet (1 m), about one-fifth of Bangladesh would be underwater. And if hundreds of thousands of people keep being displaced every year, the country will need to figure out where they can all go.

People in many other countries are facing similar dilemmas. Every year, people around the world are losing their homes to a combination of storms, flooding, drought, and other disasters caused or amped up by climate change. As the human population keeps growing and the amount of livable land keeps shrinking, there's

Flooding in Bangladesh is increasing in severity and frequency as climate change alters precipitation patterns. Some researchers estimate that 18 million Bangladeshis will be displaced by 2060.

going to be more demand for land and resources. That's going to make climate change as much of a humanitarian crisis as it is an environmental one.

Moving Around at Home

Once climate change forces people to permanently leave their homes, those people will fall into one of a few categories: internally displaced persons, migrants, or refugees. Those categories are similar, but the differences are important.

AND TO BUFFALO ROAM

In the United States, some northern cities have started planning for an eventual influx of Americans leaving extreme weather in Florida and other coastal areas or drought in the southwest.

Buffalo, New York, is a perfect example. Once one of the country's biggest cities, it's now home to fewer than half as many people as lived there in 1950. But it has the capacity to grow again, and city officials have already started talking about the city as a potential climate refuge. After Hurricane Maria (see page 60), the city welcomed about ten thousand internally displaced people from Puerto Rico. "Based on scientific research, we know that Buffalo will be a climate refuge city for centuries to come," Buffalo mayor Byron Brown said in February 2019.

Cincinnati, Ohio, another city with far fewer people than it had in the 1950s, has also started to market itself as a future destination for the internally displaced. Ohio was a popular destination for Louisianans relocating after Hurricane Katrina, and Cincinnati's 2018 green plan included using the city's location to attract those in need of new homes in the years ahead. Other Midwestern cities that have lost much of their former populations, from mining towns such as Duluth, Minnesota, to "Rust Belt" locales in Pennsylvania, Ohio, and Michigan, are potential homes for internally displaced Americans and refugees from elsewhere.

As the name suggests, internally displaced people have been forced from their homes and need to find new ones, but are searching within the same country. This status is usually less complicated than being a refugee, but it isn't entirely uncomplicated. When, say, a coastal town becomes too flooded for people to get by or desertification makes a town unlivable, people need to resettle. The first option will usually be to find somewhere else to live without moving abroad.

If the environmental damage is temporary (say, from a specific storm), many of those people will have the option of eventually moving back. But if it's permanent (say, from desertification), they're going to need to start new lives in a new town. Even in the best cases, people experience major upheaval including changing schools, finding new jobs, and becoming disconnected from at least some friends and family. Of course, most climate change impacts aren't temporary, and they're probably going to get worse.

People all around the world are already being displaced by climate impacts such as droughts, rising seas, and extreme weather. For example, after Hurricane Katrina, about one million Americans in Louisiana and other Gulf Coast states were forced to leave their homes for at least some time. A lot of those fleeing the hurricane damage had no idea at the time if New Orleans and the area around it would be able to rebuild or when that might happen. About one quarter of a million people leaving the Gulf Coast settled in the Houston metro area. By 2015, a full decade later, an estimated one hundred thousand of them still lived in the Texas city and its suburbs. While Houston (which suffered its own crisis due to Hurricane Harvey in 2017) received the biggest total, many other American cities, from Atlanta to Chicago to San Francisco, welcomed large numbers of people who lost their homes to Katrina and still serve as homes for many.

Some of the former Gulf Coast residents stayed away for positive reasons—they'd gotten comfortable in their new hometowns or found new jobs or relationships somewhere else. For others, the decision came down to concerns about returning to the Gulf Coast. Rebuilding took a lot of time and money, and many former residents didn't have enough of one or both. Some had no reason to go back. If their home was destroyed and their personal connections had all left, it wouldn't feel the same. Some areas still haven't completely recovered. And what if it happened again? Would moving back be worth the risk of losing everything a second time?

In January 2016, for the first time in history, the US federal government put money toward permanently relocating a group of Americans because of climate change. Some 80 miles (128.8 km)

The Isle de Jean Charles in Louisiana has seen almost 98 percent of its land disappear due to erosion and rising waters. Despite this, some residents remain there and are adamant that the island will not disappear. This is another case of climate change making an existing problem worse. In this case, that includes the existing network of levees and dams disrupting the flow of the delta.

southwest of New Orleans, the island Isle de Jean Charles sits in the Mississippi River delta and has lost all but about 2 percent of its territory to erosion, storms, and rising waters since the 1950s. It shrank from 22,400 acres (9,065 ha) in 1955 to 320 acres (129.5 ha) in 2016.

As part of a Department of Housing and Urban Development grant program to address climate change (which gave $1 billion to thirteen states that month), the federal government allocated $48 million to Louisiana to move the population of Isle de Jean Charles. Three years later, the state finally purchased about 515 acres

(208.4 ha) to relocate the whole remaining community. It had already dwindled to about one hundred people after various hurricanes and floods drove individuals who used to live there to find higher ground. Most of the island's resettling residents are members of the Biloxi-Chitimacha-Choctaw tribe, who had been forced onto the island from their ancestral land by the federal government in the first place. The goal of the land purchase was to make sure the island population could move together and maintain a community (though some people decided to stay as long as they can).

The residents of the Isle de Jean Charles are among the first Americans to be permanently displaced by climate change, and they certainly won't be the last. Already, more than a dozen communities throughout the country, including other Indigenous people in Alaska and the continental United States, are relocating for climate-related reasons. Experts estimate that climate-driven sea-level rise within the United States will internally displace four million to twenty million people by the end of the century. As we discussed earlier, the Gulf Coast and the Atlantic Coast are the areas most at risk, and towns in states from California to North Carolina to Maryland have already started planning for where to move people when flood damage hits a tipping point.

It wasn't unusual to hear people at the time refer to the people leaving New Orleans post-Katrina or the Isle de Jean Charles evacuees as "refugees," but that term wasn't technically accurate. Refugees, by definition, are people forced to leave their home country. Because they're staying in the same country, internally displaced people are still citizens, with all the same guaranteed rights they had before. When floods force someone to move from Florida to North Dakota—or Venice to central Italy, or Amsterdam to the southern Netherlands, or Barbuda to the island of Antigua—their life will change a lot, but their legal status won't. They rarely face travel restrictions. They are eligible to get a job, rent or buy a home, and get whatever healthcare or other benefits their government policies offer.

While they have the legal right to do all those things, whether they can afford to do them is a different question. In the United States, 40 percent of households don't have at least four hundred dollars set aside for emergencies. And while worldwide poverty has generally improved during the past century, the World Bank estimates that about one of every ten people around the world still lives on less than two dollars a day.

As many of the examples from earlier chapters demonstrate, poor people are generally more at risk from climate change. In the words of a 2019 UN report, thanks to climate change, poor people around the world will often be forced to "choose between starvation and migration" as climate change keeps spreading drought. During the three years before that report came out, an average of twenty-six million people around the world already needed to relocate every year because of natural disasters. That's an average of at least one person every second and a much larger figure than the

INTERNAL V. EXTERNAL

Numbers for how many people are internally displaced or in need of refugee protection are always estimates, but the United Nations tracks them. By early 2020, the UN estimated that 41.3 million people around the world were currently internally displaced and cited Syria, Colombia, Somalia, and the Democratic Republic of the Congo among the countries with the largest displaced populations. The UN also considered about 25.9 million people to be refugees, with two-thirds of them coming from Syria, Somalia, South Sudan, Afghanistan, and Myanmar. (Many of those refugees from Myanmar have gone to Bangladesh, adding to the number of people at risk from rising seas there.)

number of people displaced by wars and other violence during the same time. Like the things causing it, environmental migration isn't a future problem.

Refuges for Refugees

When climate change forces people to leave their homeland and enter a new country, non-governmental organizations, journalists, analysts, and even some world leaders will refer to these people as "climate refugees." However, this term isn't yet considered official. Because the term refugee carries legal status under international law, some government officials and other institutions have worked hard to avoid using that term for people fleeing environmental disasters, even when they fit the definition. Otherwise, these governments would be obligated to do something about it, as the vast majority of countries have signed international treaties that set rules regarding how to treat refugees. (See the sidebar on page 140.) Some governments have been open to taking in climate refugees after specific disasters, but even they don't classify them as refugees, giving themselves some leeway in how they deal with the influx of people. Consequently, few world governments have begun to move entire groups of people who have lost their homes or livelihoods to climate change.

NORTH AMERICA

Tropic of Cancer

PACIFIC OCEAN

Equator

SOUTH AMERICA

Tropic of Capricorn

Desertification or drought

Hurricanes

Deltas: at risk of extreme flooding

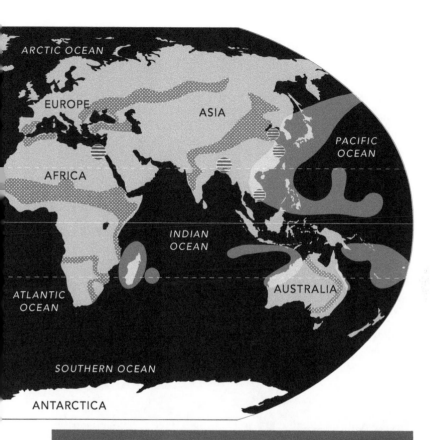

ARCTIC OCEAN

EUROPE

ASIA

PACIFIC
OCEAN

AFRICA

INDIAN
OCEAN

ATLANTIC
OCEAN

AUSTRALIA

SOUTHERN OCEAN

ANTARCTICA

This map indicates where the most pressing climate change concerns are occuring throughout the world and, therefore, where internally displaced people and climate refugees are most likely to come from.

Nor are world leaders keeping up with the speed of the climate crisis. In January 2020, the United Nations Human Rights Committee ruled for the first time that countries can't deport people to, or force them to stay in, places where their life is at risk from climate change. In that specific case—that of a man from Kiribati who was denied asylum in New Zealand and deported—the UN committee ruled that the would-be refugee's rights weren't violated.

RIGHTS FOR REFUGEES

Internationally, the legal basis for refugees comes from the 1951 United Nations Convention Relating to the Status of Refugees. It was initially a treaty about the rights of European refugees from World War II and set rules for how the countries that signed the agreement had to treat them.

But considering how many other situations were forcing people to move and how many non-Europeans that included, people realized that the 1951 treaty didn't go nearly far enough. So, most nations of the world agreed to a second treaty.

The 1967 Protocol Relating to the Status of Refugees expanded the same protections for all refugees. As of 2020, the only three countries that haven't adopted the 1967 protocol are Turkey, Madagascar, and St. Kitts and Nevis. The rest of the world is supposed to follow the rules.

As to what those rules cover? Countries are not supposed to deport refugees to places where they might be persecuted or to other countries that might then send them to those kinds of places. They're supposed to give refugees legal residency status (not necessarily citizenship) and give them a chance at a job, an education, and federal benefits. The protocol doesn't guarantee that they receive all those things but states that they're entitled to apply for them fairly and receive them if they qualify.

Also, countries are required not to punish refugees for entering without a passport, visa, or other legal documentation. (A country can reject an individual person for a specific reason, such as a criminal history, but is supposed to admit the refugee unless something like that comes up.)

But the committee's ruling also created a precedent in international law that attorneys can use on behalf of future climate refugees.

That ruling was important, and it will probably urge some countries to take the problem more seriously. On the other hand, international law is often difficult to enforce. (That's a challenge with climate agreements in general.) An enforceable worldwide standard for how to treat climate refugees may still be years away. The delay understandably hasn't stopped people who need to move in the meantime.

In 2014, Kiribati became the first country to purchase land abroad as a possible home for its climate refugees. For a little less than nine million dollars, the government in Tarawa bought about 7.7 square miles (20 sq. km) on Vanua Levu, the second-largest island in Fiji. That land is being treated as a last resort, an investment in the future, and a way to keep I-Kiribati people from emigrating somewhere else out of fear that they won't have a place to move when things go bad. For the time being, the Kiribati government has planned to use the land in Fiji to grow root vegetables and other crops to help make up for some of the food supply lost to rising seas back home.

Other countries at high risk have looked into similar land purchases. But not all have the same resources. For Tuvalu, for example, nine million dollars is more than a quarter of the country's total gross domestic product. And while Fiji was open to selling some space to Kiribati, that's hardly an approach most countries are following. It's one thing for a small island country like Kiribati to try moving a few thousand people; as more populated countries like Bangladesh lose land, relocating millions of people is a very different challenge.

Researchers have been talking about the concept of environmental refugees since at least the 1980s, and the 1990s saw several studies trying to estimate the current number of such people and how many the world could expect in the years ahead. But the issue gained a lot more traction by the early twenty-first century, and governments and scientists have been seriously studying migration from at-risk areas.

The World Bank predicted in 2019 that by the year 2050, huge numbers of people from three of the world's most at-risk regions—

sub-Saharan Africa, Latin America, and South Asia—will need to migrate. The expected number is about 143 million people. That's more than the 2020 population of Mexico and not far behind that of Russia. Only nine countries currently have more people than that. The good news is the same 2019 study pointed out that a serious reduction in carbon emissions could reduce that number by half or more. The bad news is that caveat still feels very optimistic, and plenty of less-conservative models place the number much higher.

We won't know the ultimate numbers for a while, but we do know that climate change is at least partly responsible for many of the refugees already on the move.

Already a Hot-Button Issue

Since about 2014, many thousands of people have fled the "Northern Triangle" nations of Central America: Guatemala, El Salvador, and Honduras. In most cases, they point to violence or poverty as their reasons for needing new homes. The Northern Triangle is now considered among the most dangerous parts of the world—especially for women—and every other nearby country (the United States, Mexico, Belize, Nicaragua, Costa Rica, and Panama) has seen a huge increase in the number of asylum applications from those three countries. People in the Northern Triangle are also poorer, on average, than those in other Central American nations. In Guatemala and Honduras, about 60 percent of people live below the national poverty line, while for Central America overall it's only about 30 percent.

Although climate change isn't the main driver of that migration crisis, it has contributed to why the situation has gotten so bad. The Northern Triangle makes up most of what is called the "dry corridor," a stretch of land in Central America that's already vulnerable to both drought and extreme weather.

For many Central American refugees, the attempt to move to a new country was part of a two-step relocation. Many originally lived in rural areas and worked in agriculture. A lot of farmers from the Northern Triangle grew coffee, one of the most important crops in the region (and one of the most important for office workers and students worldwide). Small farms still produce much of the coffee that comes from Central America. As recently as 2015 about 80 percent of Central American coffee came from farms less than 8 acres (3.2 ha) in size.

Unfortunately, these valuable beans have been hit hard in recent years by coffee leaf rust, a contagious fungus that gradually destroys coffee plants. One way coffee plants have evolved to survive is to grow well at higher, cooler elevations. The fungus doesn't spread as well in those conditions, and farmers usually plant coffee trees well above sea level for that reason, among others. But because those areas are now warmer than they used to be, the blight can now spread and wipe out the livelihoods of many farmers. An outbreak began around 2012, and by 2016 the leaf rust had hit seven out of every ten farms in the region, costing 1.7 million workers their coffee-farming jobs.

Add a years-long drought on top of that. Extreme heat and a lack of rainfall hit parts of the Northern Triangle in 2014 and 2015, killing about three-quarters of the bean and corn (or maize) crops, as well as thousands of cattle. Coming right after the coffee crisis, the drought made food hard to come by and reduced income even more for both workers and producers. According to the UN's Food and Agriculture Organization, years of drought killed 694,000 acres (281,000 ha) of beans and corn in Guatemala, Honduras, and El Salvador by 2018.

This is one big reason why Central Americans make up about three million of the 143 million people the World Bank expects will become climate refugees by 2050. In some interviews, Guatemalan migrants have talked about their children regularly going to bed hungry for years thanks to crop failures and wage cuts. Without being able to produce enough food and without earning enough money to purchase it, thousands of people in the Northern Triangle moved from the countryside to cities.

But moving to urban areas wasn't always a great option. Facing gang violence, widespread corruption, and other dangers in the cities, many needed to hit the road again and try to find a new country entirely. Sometimes, that meant trying to go places where they weren't always welcome.

Climate refugees face many of the same challenges as other immigrants, at a time when restrictions on migration are already becoming stricter in several key countries. This has definitely been true in the United States, where migration from Central America through Mexico has become a hot-button issue in recent years, particularly since the start of the Donald Trump administration. Starting in 2018, thousands of Central American migrants fleeing the Northern Triangle headed north through Mexico and on to the United States to apply for refugee status.

In the United States, the federal government under the Trump administration treated these migrants—who were following a legal migration process protected by international law—as if they were undocumented immigrants crossing the border illegally. Thousands were deported to Mexico or held indefinitely in cages, which sparked international outcry. Many children were separated from their parents, some for good, by the harsh and illegal US response. Others were sent back to dangerous situations in their home countries that they didn't create and will need to try migrating again or find a new way to survive.

Not all climate migrants around the world are going to face a situation as dangerous as the violence seen in the Northern Triangle cities—but the poor in the developing world are the people most likely to need new homes. And we still don't have a global plan for what to do about that.

A woman works on her drought-stricken farm in Kenya.

Feeding the World

The droughts and crop failures that help drive human migration are spreading, which affects most people on Earth to some degree. We live in a global economy, where food supplies are interconnected. In our current world, this is easy to take for granted.

Trading food has been a key part of international commerce for most of recorded history. A few millennia ago, people risked their lives to bring spices such as ginger and cinnamon from Asia to Europe or to ship grain or olive oil across the Mediterranean. While apple pie is an American cliché, North America didn't have apples until the 1600s. Tobacco didn't get to Europe until Christopher Columbus brought some back from North America, and coffee didn't get there for more than a century after that. People in the twenty-first century rely even more heavily on food produced outside their countries' borders. The United States is both the biggest importer and biggest exporter of food when it comes to total amount; China, Germany, and Japan also rank near the top of both lists. But even many of the world's poorest countries rely on imported grain or vegetables to prevent food shortages. In other words, crop damage impacts countries around the world.

Drought isn't the only climate-driven problem facing the world's food supply. Many other climate-change effects are also destructive to crops. Extreme rain, storms, and coastal flooding all damage soil. So do desertification and deforestation. And because the world population is still growing and all those extra people will need food, pressure on the areas that can still grow it is increasing.

A 2018 United Nations study warned that at the rate things are going, by 2050 some regions could see their crop yields (the amount of crops actually harvested, not just planted) fall by as much as half. The International Food Policy Research Institute (IFPR), using a more aggressive model, warned that global yields of several crucial crops—including corn, wheat, rice, and potatoes—could fall dramatically by 2050. The IFPR predicted that corn could be the hardest hit of those crops, with global yield falling by nearly one-quarter from its 2000 level. Whether those troubling predictions come to pass depends on whether we can rein in climate change and whether farmers are able to diversify their crops and keep soil from degrading (or even better, if they can adopt more sustainable practices and replenish the soil).

As with any other climate impact, not all regions will suffer similarly. Some areas farther north will have longer growing seasons as the planet keeps warming and will be able to grow crops that couldn't survive there earlier. These areas include parts of northern Europe, the northern United States, and southern Canada—places still far enough south of the Arctic that crops can thrive, but far enough from the tropics to avoid the deadly heat there. But even this won't make up for the falling crop yields in warmer places. In the United States, for example, Illinois and Indiana are already seeing annual corn yields fall by up to 8 percent, while yields in Iowa and Minnesota have grown by only about 2.8 percent.

Climate Change Causing Conflict

The first genocide of the twenty-first century happened in the Darfur region of western Sudan. Government-funded airplanes flew over farming villages, dropping bombs and destroying homes, killing those who couldn't get away. Militias called Janjaweed, with large groups of fighters riding horses or camels, then attacked villagers in that part of the country—raping, torturing, and murdering thousands of people.

These attacks started in 2003 and went on for years. Though there is no official death toll, estimates range from two hundred thousand to half a million people murdered. World leaders including US president George W. Bush and officials like Luis Moreno-Ocampo, the chief prosecutor of the International Criminal Court, publicly called what happened in Darfur a genocide, but many countries didn't intervene militarily, sanction Sudan, or punish the guilty as required by international agreements about genocide. The attacks internally displaced two million people, forcing them out of their homes and into refugee camps. It wasn't until July 2007 that the United Nations sent a peacekeeping force to try to control the violence. It took another year before the International Criminal Court charged Sudan's president with war crimes (though he stayed in office until a coup removed him in 2019).

Ethnic tension was definitely a big factor in the violence, with the attackers, mostly nomads who identified as Arab, killing Black farmers who identified as African. (How those terms are used in Sudan is too complicated to get into here, but the important part is the two groups viewed each other as different peoples.) Civil war

PREPPING FOR CLIMATE COMMOTION

Unlike the amount of mass a glacier has lost to melting or how many degrees the temperature has gone up, the impact of climate change on war and other conflicts is hard to measure. But it's something US defense and foreign policy experts have been worrying about for a long time.

As far back as the Ronald Reagan administration (1981–1989), government defense planners warned that climate change was going to become a national security issue for the United States. By 1991, the annual National Security Strategy, under president George H. W. Bush, briefly mentioned climate change among a series of environmental issues that would play a role in post–Cold War security.

But in the 2000s, the US military and intelligence communities were constantly producing warnings. A 2003 Pentagon-commissioned report warned that "violence and disruption stemming from the stresses created by abrupt changes in the climate pose a different type of threat to national security than we are accustomed to today." The *Quadrennial Defense Review* from the US Department of Defense (DOD) has issued similar warnings every four years, joining a pile of similar reports. Other countries' experts have done the same.

So, even when President George W. Bush (the second President Bush) routinely insisted climate change wasn't settled science (when it was) or when Donald Trump claimed it was a "hoax" (which it wasn't), their own defense departments were actively planning for climate-change contingencies. A 2019 DOD study warned that seventy-nine US military properties were already vulnerable to climate change, and the intelligence community's 2019 *Worldwide Threat Assessment* addressed the "growing threat" of "increased natural disaster, refugee flows, and conflicts over basic resources." It's not unlike how oil companies' scientists were warning about the dangers of emissions while the companies' leaders were denying them.

It's worth pointing out that the US military remains a major source of greenhouse gas emissions even as its scientists warn against them. Exact numbers are hard to come by, but one 2019 *Newsweek* study estimated the US military alone would rank forty-seventh among all countries in annual greenhouse emissions.

was another factor, as rebel groups based in Darfur had taken up arms against the central government in Khartoum, though the government's backing of a genocidal response against civilians was obviously evil and extreme. But even as the genocide was in progress, another underlying cause started to get international news coverage—the role of climate change in kicking off the crisis.

As we've discussed throughout this book, the extreme heat and odd weather patterns brought about by climate change create shortages of natural resources. We've talked about how prolonged droughts can lead to water shortages and force cities to ration the most important resources on the planet. Plus, droughts and desertification can make certain areas uninhabitable, leaving less land to go around. In severe cases, people will fight over limited resources, making war and conflict indirect results of climate change.

In the case of Darfur, several of these factors played parts in the conflict. By 2003, the area was getting about 30 percent less rain than it did in the 1960s, and the growing Sahara had pushed south about 60 miles (96.6 km) in that time. In 1984, the region went through a deadly extended famine, during which many herdsmen lost their pastures to drought and became nomads. Conflicts over water or land sometimes turned violent. While many have discussed the 2003 war and genocide as a conflict between "Africans" and "Arabs," it was also between settled farmers and the nomadic herdsmen who strategically drove them from their land.

Climate change isn't exclusively to blame for any of these conflicts. Most of the time, an underlying situation was already there, and it's possible any range of factors could have lit the proverbial match. Even so, experts warn that climate change will make fights over resources more common and more violent. It's why the DoD calls climate change a threat multiplier, something that makes situations that are already volatile more likely to explode. Numerous researchers are working on modeling the possibility of civil unrest exacerbated by the effects of climate change.

For another example, several factors caused the Syrian Civil War that started in 2011, but the war also began after a record drought, the worst the region faced in nine centuries. In just a few years, the country's rural areas lost most of their crops, which made food really expensive and led to widespread malnutrition and disease. As in the Northern Triangle, farmers who lost their crops, livestock, and jobs moved to cities in huge numbers. Protests in Syria during the Arab Spring of 2011—a series of large protests that toppled governments in countries such as Tunisia and Egypt—and the Syrian government's brutal retaliation against them escalated into a horrifying civil war that has killed more than a quarter of a million people. Climate change wasn't the sole cause of the war, but the poverty, hunger, displacement, and social unrest that drove it were amplified by the severe drought.

Extremist groups have also used climate impacts as opportunities to recruit from populations who lost their livelihoods and homes. The rise of the Islamic State of Iraq and al-Sham, or ISIS (also known as ISIL, if you substitute "the Levant" for "al-Sham"), would not have been possible without the combination of massive drought and civil strife in Syria. Both there and in Iraq, which also suffered a severe drought and subsequent poverty, ISIS was able to use the environmental crisis to its advantage, offering money to recruit ruined farmers. With most of its fighters coming from rural areas, the terrorist group took over more than 17,000 square miles (44,000 sq. km) by its 2014 peak. Though ISIS lost most of that land in the next two to three years, it increased the odds of future conflicts over resources in the region by destroying irrigation systems, poisoning wells, and looting resources in areas it conquered. These acts were done both to harm potential ISIS enemies nearby and to make more people desperate enough to consider the group's recruitment offers. With all that damage, the region will no doubt suffer badly in the next major drought or other catastrophe.

The Nigerian extremist group Boko Haram (which translates roughly to "books are forbidden" or "Western education is forbidden") is probably best known for the 2014 mass kidnapping of 276 teenage girls from school and its brutal attacks on public schools and students, such as trapping everyone inside and then burning classrooms or dormitories. It does most of its recruiting in the areas around Lake Chad, a huge lake in the northern part of sub-Saharan Africa that spreads across parts of Chad, Niger, Nigeria, and Cameroon. Drought and land degradation there have been a recruiting boon as, like ISIS, the group was able to offer money and work to those desperate for it.

As you can imagine, many people want to get far away from civil war, extremist groups, and other violence. Europe has seen huge numbers of migrants from the Middle East and North Africa fleeing these kinds of conflicts. In 2015 alone, more than one million people from those regions sought asylum in the European Union, forcing those countries to deal with a major domino effect of climate change.

All these pressures will change the world in the years ahead. Whether what we're doing to the climate causes a specific instance of war, migration, hunger, and other problems or just makes them worse, world leaders—and all of us, on some level—need to plan and adapt for new crises. In the last chapter, we'll talk about some of the possibilities for limiting the damage caused by climate change.

CHAPTER 7

SO, WHAT CAN WE DO ABOUT IT?

We need to get angry and understand what is at stake.
And then we need to transform that anger into action
and to stand together united and just never give up.

—teenage climate activist Greta Thunberg in 2019,
the year *Time* magazine named her Person of the Year

BY NOW, YOU'VE seen many examples of how climate change is already a worldwide problem that will affect everyone on Earth. This may not be fair, but that's the way it is. The efforts of people from all over the world are needed to solve this problem.

The environment, after all, is something we all share. How people use common resources such as water, air, and land affects how others can use those resources. And the same is true for how they pollute them. What's happening to the climate is happening to everyone, whether they personally emit a lot of greenhouse gases or do their part to take care of their environment.

People in Kiribati or the Maldives aren't the ones causing climate change, and their societies were able to get by for centuries before the current crisis. And animals such as polar bears, sea turtles, and koalas didn't cause the problems that are now putting their species at risk.

And let's be clear. Relatively few people are actively trying to make climate change worse, and most people are at least open to making more sustainable choices. So, we need both individual and collective action if we're going to limit the damage from climate change. There are a range of actions people can take, and they're not mutually exclusive. We can try many or all of them, and we'll walk through a few examples.

Cutting Back for Good

The deadly COVID-19 pandemic that exploded early in 2020 was a horrible situation. It forced billions of people worldwide to limit their social interactions and lock down in their homes, infected and killed millions of people, destroyed businesses and jobs, and made life more difficult for nearly everybody. If there was one tiny bit of good news in the chaos, however, it was that the situation temporarily reduced human impacts on the environment. Pollution briefly fell for the first time in years with better air quality, fewer industrial activities that damaged the environment, and city skylines that used to be covered in smog suddenly visible both on the ground and from above.

Obviously, a global pandemic is not how we want to reduce emissions, but the situation proved that it's possible to cut environmental damage in an emergency and see results. Daily global carbon emissions worldwide fell by 17 percent in April 2020, though the temporary nature of the lockdowns meant the improvement didn't last long. We can move in that direction with a planned strategy. Climate change is an emergency in its own right. This is why international agreements on reducing emissions are so important.

Students use recycled materials to make art projects. In addition to art, people can make jewelry, clothing, blankets, writing utensils, paper, and more from recycled materials. Metals and certain glasses can be continuously recycled, while paper can be recycled about five to seven times. Plastics, however, can only be recycled once or twice before having to be converted to fabrics or another non-recyclable material (and some plastics can't be recycled at all).

Here is where the idea of a carbon footprint comes in. Any human entity—an individual person, a business, a country—requires a certain amount of carbon to support its activity. A carbon footprint gets bigger not only from directly burning fossil fuels (say, driving a car or cooking on a gas-powered stove), but also indirect actions that rely on burning fuel (say, buying a gift shipped by airmail or eating a hamburger, as the processes to produce and transport those things cause substantial emissions). Everyone has a carbon footprint; it's unavoidable. The key is shrinking the world's combined carbon footprint enough to be sustainable, a level that can safely exist as is without causing long-term harm to the environment. Individuals must do their part to reduce their emissions, but also—and even more importantly—governments and companies must do the same.

The most direct way to cut greenhouse gas emissions is to produce less of them. Seems simple, right? There are two main ways to do this: cut back on activities that burn fossil fuels or otherwise produce greenhouse gases and replace devices or processes that use fossil fuels with ones that either don't use them or do so in a more energy-efficient way.

Let's look at transportation as one example, since that sector is one of the biggest sources of greenhouse gas emissions. According to the EPA's most recent data, transportation makes up the biggest share of Americans' greenhouse gas emissions (28 percent), just ahead of energy (27 percent). This category includes not only the emissions from cars, but those from airplanes, trains, buses, boats, and ships. And transportation is a major source of worldwide emissions as well.

Vehicles and transportation are major sources of emissions because they're so common. But there are a lot of opportunities to both cut back on their use and make them more efficient. Public transit is part of the solution. One train or bus carrying hundreds of people to and from work every day has a much lower carbon footprint than the same number of people traveling in separate vehicles. Carpooling when possible, combining a series of errands into one trip, and walking or biking more often are all good options for how you can cut back on emissions in your personal life. Changing the way we work is another piece of the puzzle. The COVID-19 pandemic showed that many jobs could be done from home at least some of the time, and companies dramatically cut back on the amount of air travel for business. If people can cut down on traveling or commuting when they don't need to do it, they will reduce even more emissions.

FIXING A HOLE

Successful international action to get rid of dangerous chemicals in the atmosphere isn't just possible; it's happened before. In the 1980s, governments took action to deal with damage to Earth's ozone layer.

Located in the lower part of the stratosphere, the ozone layer helps absorb most of the sun's ultraviolet (UV) radiation—more than 97 percent of it. In small doses, UV rays give us vitamin D and suntans. In larger amounts, they bring serious health problems like severe burns and skin cancer.

Chlorofluorocarbons (or CFCs), sometimes called Freon after one specific brand, were common refrigerants also used in aerosol cans and air conditioners. First manufactured in the 1930s, these chemicals were effective at keeping things cold, but they also interacted with and damaged the ozone layer. Scientists figured this out by the 1970s, and most of the world caught on in 1985 when a huge hole in the ozone layer over Antarctica became too obvious to ignore. The hole—technically more of a dramatic thinning of the layer—was big enough that most of that continent was under it, and other nearby areas also lost protection from the sun.

Without international action, the EPA estimates, the hole could have covered the entire planet by 2050. And just in the United States, there would be up to three hundred million more cases of skin cancer. But in 1987, the United Nations created the Montreal Protocol on Substances that Deplete the Ozone Layer. This was a treaty to stop production of CFCs. Signed by all 197 recognized nations, it took effect in 1989. It banned the production and use of certain CFCs and phased out others more gradually.

The treaty has been pretty successful too, as the world has phased out about 99 percent of the ozone-harming chemicals banned under the treaty. It's also helped on the climate-change front, as some of the chemicals that destroy ozone are themselves powerful greenhouse gases. Some estimates of what would have happened without the agreement predict the planet would have warmed by about 4.5°F (2.5°C).

The hole is still there. In fact, it's one of the reasons why Australia and New Zealand, the countries closest to the hole, have the world's highest rates of the skin cancer melanoma. But it has begun to heal. Concerted effort across the globe prevented it from getting much worse, and that's an important win—one that shows international action can work.

Replacing high-emission vehicles with more efficient ones is another important option. Improving vehicles' efficiency requires better technology to increase the distance the vehicle can travel on the same amount of fuel. In the United States, mileage standards originated in the 1970s as a response to spikes in the price of oil. When Congress passed the Corporate Average Fuel Economy (CAFE) standards in 1975, it set a goal of passenger cars getting 27.5 miles per gallon (11.7 km/L) in the next ten years. That was more than twice as efficient as cars were at the time, and the auto industry hit that goal (though there was a short period in the late 1980s when the Reagan administration slightly lowered the requirement). During the Barack Obama administration, the US federal government set a goal for cars to reach 54.5 miles per gallon (23.2 km/L) by 2025, with benchmarks of 5-percent improvements every year along the way so that new vehicles would always be more efficient than the previous year's cars. Standards for trucks require less efficiency but have still improved since the 1970s. The European Union has even higher standards, requiring manufacturers' vehicles to average 57 miles per gallon (24.2 km/L) by 2021 and a goal of 92 miles per gallon (39.1 km/L) by 2030.

There's another benefit of fuel efficiency: it costs drivers less money. If your car gets fifty-four miles per gallon instead of twenty-six, you're filling up only half as often even if you drive the same amount. That's big savings—especially if you drive a lot—and a good reason for even the stubbornest of science deniers to want a greener alternative.

Many of the more efficient cars still use gasoline engines that are designed to do more with less fuel. Others are hybrids, meaning they also have at least one electric motor and a battery, which are powered by recapturing some of the energy lost by braking. More recently, some companies have developed electric plug-in cars, which use an electric motor and battery without directly relying on gasoline. Hybrids and electrics tend to be more expensive than gasoline-powered cars, but like most technology, they'll get cheaper as they become more popular and manufacturers make more of them. This concept is known as economies of scale and it's crucial for addressing climate change.

Government and business decisions to make efficient vehicles more affordable and to focus on greener technology are important parts of the equation. For instance, although cars are more plentiful than other vehicles, other forms of transportation, including those controlled by government agencies, can become more efficient too. For more than a decade, major cities including New York, Seattle, and Washington, DC, have been switching to hybrid buses that run on a mix of diesel fuel and electricity. Other cities around the world have experimented with hybrid train technologies. Planes are the big

As the materials and processes required to manufacture an electric car cost less and less, more car companies are offering hybrid and electric vehicles at affordable prices. Switching to an electric vehicle can save purchasers a lot of money in the long run and decreases their carbon footprint.

wild card because they burn so much more fuel than cars do. Making them more efficient is going to be crucial.

Home appliances are another good example of how technology can become greener. Electricity is the biggest source of emissions worldwide and a close second in the United States, so if every household used energy-efficient appliances, emissions could decrease substantially. Think about something as simple as the lightbulb. In recent years, consumers have been able to replace their regular bulbs with more efficient options, such as compact fluorescent lamps (CFLs) and light-emitting diodes (LEDs). While the old-fashioned incandescent bulbs waste most of their energy as heat rather than light, the new kinds don't waste energy that way. They also last longer and cost less to use. According to the US Department of Energy, CFLs and LEDs use 25–80 percent less energy than incandescent bulbs and last up to twenty-five times longer. For bigger home appliances such as refrigerators, air conditioners, and ovens, the US government provides guidance on which ones conserve energy. Since 1992, the EPA and Department of Energy have certified appliances with an "Energy Star" rating if they hit certain efficiency targets, and other countries have similar programs that let shoppers know which appliances are more efficient. Cheaper to use, less energy, lower emissions. A win all around.

Animal agriculture is another big cause of emissions. Studies by scientists and intergovernmental organizations, such as the IPCC, attribute anywhere from 14–18 percent of greenhouse gas emissions to the meat and dairy industries. But those numbers understate the problem because methane makes up a lot of those food-production emissions. And, as we discussed back in chapter 1, methane is many times more damaging than carbon when it comes to climate change. The IPCC estimates that a little more than one-quarter of global emissions come from agriculture and more from meat than any other product. (See the sidebar on page 161.) A 2018 Oxford University study called a vegan diet the "single biggest way" for individuals to reduce their carbon footprints, estimating it could reduce them by about 73 percent. Even just cutting back on meat consumption would make a difference on your personal carbon footprint, and trends such as "meatless Mondays" have gotten more popular as ways to reduce emissions. Plant-based meat alternatives have become more popular too, with sales growing about 40 percent between 2017 and 2019. Many plant-based options have carbon footprints about 10–20 percent of the size of beef's. Even major chains such as Burger King, KFC, and Taco Bell have added them to their menus.

Using less energy as a planet is challenging, with technology becoming more important in people's lives around the globe. But we have to do it, and it will need to involve a mix of all the things we're talking about. We can all take common-sense steps such as turning off lights or appliances when not in use, recycling, and buying efficient devices when possible. If everyone does that, it will help a lot, but that alone isn't enough. There are other important steps that need to be taken on a larger scale.

Something Better Comes Along

Another key part of solving the climate crisis is replacing fossil fuels with cleaner, renewable energy options. The faster that happens the better. The good thing is there are a range of renewable options already in use. The point isn't to pick one of these instead of the others; a sustainable future is going to need a mix of energy sources.

Solar energy involves capturing energy from the sun, an energy resource we've been using since the dawn of humanity. Converting the sun's energy to power devices is not exactly a new technology. In the 1830s, British astronomer John Herschel famously used an oven that ran on solar power. There are a few different kinds of solar cells, but they all work by capturing energy directly from the sun and converting it into electricity.

ANOTHER GASSY PROBLEM

Animal agriculture is one of the biggest contributors to climate change because of methane. Cattle and other ruminants (including bison, sheep, and goats) produce a lot of the greenhouse gas because of the way microorganisms in their digestive systems break down and ferment food to create methane. And the animals then release it.

Burping releases most of that methane, but an animal's other end can too, and people who don't understand science (or have an anti-science agenda) have spent years giggling about cow farts contributing to climate change.

Domestic cows have been around in one form or another for more than ten thousand years and have been expelling gas that whole time. The reason cows are contributing so much to climate change is that farming practices have completely changed in the past few decades—for the worse.

The kinds of farms most of us saw in our children's books, with cows grazing freely on large stretches of green grass, haven't been the norm in numerous countries (including the United States) for a long time. Instead, animals are often packed into high-density feedlots on factory farms, sometimes packed in so tightly that they can barely move. They're also usually fed a mass-produced diet, without the variety of natural grazing. This treatment is bad for other reasons—it almost defines terms like inhumane and animal cruelty, and spreads diseases that humans can catch. It also results in many more animals producing much more methane. All those animals produce a lot of waste (what's known as unmanaged waste), and factory farms often dispose of it by leaving it to decompose in pits, a practice that produces even more methane thanks to bacteria.

Overall, meat production worldwide has more than quadrupled in the past fifty years, and the United States alone produces more than 100 billion pounds (45 billion kg) of meat per year. That's not natural, and it's one of the reasons why climate scientists strongly advocate switching to a plant-based diet or at least dramatically reducing the amount of meat people consume. This would also reduce the demand for factory-farmed animals and the amount of land devoted to that industry.

You might be familiar with solar-powered calculators, which work because of the same solar-cell technology. But larger solar panels, which combine many solar cells, can heat an entire house. While direct sunlight isn't as available at night, battery-storage technology makes solar energy possible during those hours. The solar process doesn't create waste, so it's really clean and cheap, apart from the cost of the panels themselves. Solar alone isn't going to replace all fossil fuels, as powering Earth would need a massive surface area of panels—millions of acres. Sunlight also isn't consistent enough to be a sole source, but it should be a big part of the mix.

Hydropower comes from turbines that capture kinetic energy from falling water and convert it into electricity. The turbines are usually built at large dams, though underwater turbines can capture energy from the movement of tides and currents. Every US state produces at least some hydropower, with most of it coming from Western states via locations such as the Hoover Dam and Grand Coulee Dam.

Wind power captures energy from the movement of air. Wind farms can consist of many turbines in the same area, but other locations might use just one or two turbines. The pinwheel-like turbines are becoming a more common sight; thousands are built and put into use every year.

Geothermal power is collected from heat stored underground in the magma layer that powers geysers, hot springs, and volcanoes. The heat is usually captured by drilling cold water into the ground. The water interacts with the magma layer and comes back as steam that can be used as power.

Biofuels come from formerly living sources, including crops, agricultural waste, or even algae. Corn-based ethanol is one biofuel used widely in the United States and is often included in gasoline as a way to reduce emissions. Sugarcane ethanol is another example. Switchgrass—a prairie grass with long roots that grows on land unsuitable for most agriculture—is one of the lowest-emission biofuels. All these biofuels produce less carbon than fossil fuels, but still produce more than other renewables, such as solar, wind, or hydropower. They can be useful but need to be created in a sustainable way. Using waste from existing crops is helpful, but clearing forests to grow biofuel crops would be counterproductive.

Another alternative energy source that some people champion is nuclear energy, which is far more of a mixed bag than the other examples. On the one hand, it produces little greenhouse gas pollution and is relatively cheap to produce once a nuclear power plant is built and operational. But it's also risky, as the radiation it produces is deadly. High-profile accidents such as the nuclear meltdowns at Three Mile Island in the United States in 1979, Chernobyl in what's now Ukraine in 1986 (then part of the Soviet Union), and Fukushima in Japan in 2012 are just a few examples of what can go wrong. Nuclear energy also produces waste that remains radioactive for years (in some cases, hundreds or thousands of years), and that waste needs to go somewhere. A common disposal technique is to bury it deep underground, where there's always a risk that it could eventually leak into the soil and groundwater. The other alternative energy options don't carry that level of potential danger.

One challenge for a lot of people and governments has been the cost of switching to greener energy. Like most technologies, green technologies can be expensive to adopt early but get cheaper as they become more common, and their prices will only continue to fall. It's a positive example of the feedback loop concept we talked about earlier.

Americans might think of oil and gas as cheap. When you pull up to a gas station to fill up your car, it generally costs anywhere from $2 to $4 per gallon, depending on where you live and the often-changing market price of oil. But you're actually paying a lot more for it because the federal government and some state governments use taxpayers' money to heavily subsidize the oil and gas industry to the tune of billions of dollars per year. Gasoline in Europe and some other countries often costs twice that much because governments there don't subsidize it as heavily. The deceptively cheap cost of gas in the United States is a specific government action that has had an impact on individuals' carbon footprints, and government action to subsidize renewables instead could make those cheaper and more attractive to drivers than fossil fuels.

So, what do renewables cost? According to 2019 evaluations by the intergovernmental International Renewable Agency, which includes 160 member countries: hydropower costs an average of a nickel per kilowatt hour; solar, biofuels, onshore wind, and geothermal come in at less than a dime; and offshore wind costs about thirteen cents. Fossil fuels run between a nickel and fifteen cents per kilowatt hour. In other words, renewables have already gotten competitive with fossil fuels, and economies of scale mean they could get even cheaper as we adopt them in more places.

Switching to green energy isn't some impossible future goal. Even big polluters like the United States, China, and India have increased their use of green energy, and renewable sources make up a solid chunk of those nations' energy mixes. Meanwhile, a number of smaller countries have already switched to renewable sources for much of their energy needs.

New Zealand, for example, gets more than half of its electricity from hydropower and more than 80 percent from renewables combined. Hydropower has been produced there for more than a century. More than 40 percent of the country's total energy is renewable, and the government is actively working to increase that percentage.

Uruguay has been a leader in South America, taking less than a decade to shift most of its energy to renewables. By 2015, renewable energy (a mix of hydro, wind, solar, and biofuels) was responsible for about 95 percent of the country's electricity, and about 55 percent of its total energy came from renewable sources. By 2021, Uruguay was on target to hit its goal of being totally carbon neutral by 2030. This is a serious success story.

Per capita, no country has gone greener than Iceland, which has used its natural position as an island full of volcanoes to its advantage. Nearly all of the country's energy now comes from either geothermal or hydropower, and about 90 percent of Icelandic homes get their heat directly from geothermal energy. Just a generation ago, the country got most of its energy from imported fossil fuels, but made converting to green energy a national priority. It isn't alone. European countries in general have cut emissions dramatically in the past three decades; from 1990 to 2018, European Union emissions fell by more than 23 percent overall. The United States has lagged behind those countries in adopting greener energy, but even it managed to double its use of renewable power between 2000 and 2018, and renewables made up about 17 percent of the US energy mix by 2019.

There's another benefit to renewable energy right there in the name: renewability. Fossil fuels have a limited supply. Once oil or coal or natural gas is burned, it doesn't come back (at least not without millions of years of conditions like those that formed it in the first place). Environmentally destructive practices like ANWR drilling (see pages 118–119), underground fracking, and mountaintop coal mining (which means literally blowing up a mountaintop to get to the coal underneath) are destructive on their own—even if you don't count the environmental damage caused by the burning of the oil, gas, and coal they produce. Once the extracted materials are burned up, they're gone. It's a one-time deal.

Renewable energy, by definition, comes from sources that don't run out. The sun will burn out someday in the distant future— by which time life on Earth will be wiped out anyway—but it is going to be around for millennia to come. Wind and water can shift because of weather patterns and can vary depending on the season or location, but globally we're not going to run out of either.

As with fuel efficiency, there are advantages to greener energy besides the climate impact. For one thing, these technologies are expanding, making them a good source of future jobs and offering economic opportunities for countries and businesses that lead the way. The non-nuclear options produce little to no waste. And the lack of pollution is better for people's health, as air pollution is

Oil drilling in the Arctic National Wildlife Refuge would put already threatened wildlife at further risk of population decline.

linked to diseases from asthma to bronchitis to cancer. Developing more renewable sites will mean devoting more facilities and money to those technologies, but that's not much of a downside compared to the benefits.

Undoing Damage

Cutting emissions and switching to green energy are important steps, but they only limit further damage. They don't address the carbon and other greenhouse gas pollution that's already in the atmosphere and oceans. Even if emission reduction picks up quickly—and as of 2021 it's well behind the pace mandated by international agreements—we'll still be adding to the greenhouse gas problem, just more slowly. This is where carbon sinks—natural resources that take some greenhouse gases out of the atmosphere—and carbon offsets—actions people can take to make up for the carbon emitted by their activities—come in.

We talked earlier about the importance of rain forests in managing the amount of carbon in the atmosphere (see page 84). For the same reasons, planting trees, grasses, and other plants is a valuable hedge (pardon the pun) against the climate crisis. A 2019 study published in the journal *Science* suggested that planting 2.2 billion acres (900 million ha) of additional trees around the world—a number the planet's current available space can support—has the potential to reduce the extra carbon pumped into the atmosphere between the Industrial Revolution and today by as much as two-thirds. Even if we did that tomorrow, of course, it would take years before the trees grew to maturity. We also need to protect the forests we already have from deforestation by enacting tougher restrictions on logging and clearing forests for agriculture worldwide.

Sustainable farming is important too. Vegetable and grain crops rely on carbon dioxide and help remove some of it from the atmosphere. Sustainable practices that keep soil rich both prevent the kind of erosion seen in disasters like the Dust Bowl (see page 72) and allow soil to soak up carbon itself. Soil is the world's second-largest carbon store, after the oceans. We can actively improve soil by planting cover crops, which are grown for the express purpose of covering the soil, rather than for harvesting. Farmers can plant cover crops in the season after they harvest their cash crops, or they can plant cover and cash crops side by side. Oats, cereal rye, and certain radishes are just a few examples of cover crops that help both reduce soil erosion and increase soil fertility—both things that are important for soil's ability to serve as a carbon sink.

Cities have options too. Rooftop gardens have been a popular way to add carbon sinks in urban areas, and some cities offer homeowners financial incentives to grow them. Efforts are underway to develop or popularize paving materials that will help roads and sidewalks serve as carbon sinks. Engineers have even created machines that can pull carbon and other greenhouse gases from the atmosphere. More carbon-removal technology is being developed, and the scale of the crisis means we're going to need it. And improving soil and using trees and grassland are good for the environment in general, not just for reducing carbon footprints.

Big Problems Require Big Action

All the actions individual people can take are great. If we release less carbon into the atmosphere or take other steps to improve the environment, that's always a positive. And the more people we can get to join in, the better. If everyone gets a few people they know to reduce their carbon footprints, that adds up to a lot of help.

The reality, though, is that because action to address climate change was delayed so long, addressing it now takes more dramatic and faster solutions than it would have taken thirty or forty years ago. Preventing the worst effects of climate change requires large-scale action, and some of the solutions explained in this chapter—from

One easy way to help local wildlife is to participate in volunteer programs or activities that clean up pollution in your area. Some scientists also enlist volunteers to help gather data through citizen science programs.

mileage standards to funding green technology to protecting land from oil drilling—need national or international efforts involving governments and businesses. In addition to making changes in your personal lifestyle, one of the most important things you can do is advocate for action.

One option is participating in, or even organizing, boycotts of specific companies or products with environmentally destructive practices or effects. Consumer activism can work, and there's a long history of successful boycotts and awareness campaigns—such as "name and shame" campaigns that let more people know about the unsavory practices of certain companies or politicians. It's helpful to avoid businesses that don't take action but also to reward those that do. It would be nice if all business owners cared about social responsibility, but even those who don't will make greener choices if they see a financial benefit to them.

Another option is to participate in climate strikes or marches, like those organized in 2019 by activist organizations 350.org and the Sunrise Movement, to call attention to climate change and other environmental issues. No matter your specific interests, your location, or your age, there are a lot of organizations you can work

Greta Thunberg, a teenage environmental activist from Sweden, challenged world leaders at the UN in 2019 to make changes so that their children could live in a world that hadn't been ravaged by climate change. Before speaking at the UN, she organized school strikes in which students left school to protest climate change.

with to mobilize on climate issues. (These and other organizations are listed in the Further Information section beginning on page 183.)

Climate organizations also take political action. Some have organized letter-writing campaigns to government officials and cabinet departments. Activists have protested against the Keystone oil pipeline that fossil fuel companies want to build from Canada to the United States and the ANWR drilling we talked about in chapter 5. Some, including the youth-led Sunrise Movement, endorsed climate-friendly political candidates at the federal and state levels in both the 2018 and 2020 elections, in the primaries and the general elections. Even if you cannot vote yet, you can encourage those who can to elect candidates with science-based views on climate change and plans to address the climate crisis.

Ultimately, US and international political leaders will have to take action if we're going to get climate change under control. As we've discussed, climate denial is a conspiracy theory, and anyone who endorses it has no business in a government job. If one of your representatives—from a president or prime minister down to a member of the local school board—falls into that category, they need to be voted out. They either don't understand the science, are willfully ignorant about one of the most important issues of the day, are knowingly lying about it, or are bought and paid for by someone who wants them to deny reality. None of those is a good quality in a leader, and we can do better. Climate denial needs to be treated as the unacceptable and dangerous position it is.

While in the United States, it is mostly Republicans who still deny climate change or insist it's too late to do anything, they don't have to do this. Republicans a generation ago proposed ideas like cap-and-trade programs (which would place limits on how much fossil fuel polluters could burn and let them sell those amounts to one another) and emission credits. These were ways to develop market-based solutions to climate change. Conservative parties in many other

THE LATEST NONSENSE

Now that evidence of climate change has gotten too obvious for all but the most uninformed or dishonest people to ignore, climate deniers have switched to some new arguments. One weirdly popular one amounts to arguing, "Oh well, since it's already happening, it's too late to stop it. So, we might as well keep doing what we're doing."

If you hear anyone say this, call it out.

There's a good chance that the tipping point of when climate change can't be stopped (or different tipping points for specific climate problems) will only become obvious after the fact. Like an economic recession or the fall of an empire, it's usually a lot easier to diagnose the point of no return once things have fallen apart than it is beforehand. Even if we pass such a point and it becomes too late to stop a specific climate outcome, we can still stop others and limit the damage.

When a fire destroys some houses, the fire department doesn't just give up and let it burn the others nearby. There's value in preventing a bad outcome from becoming a worse one, and the same is true for climate change.

Even if it turns out to be too late to stop some of the effects of climate change, that's no reason not to try stopping the rest or stopping the existing effects from getting worse. We have to live with the outcome either way.

countries have elected plenty of science-supporting leaders. Whether it's a primary within a party or a general election between parties, there are many opportunities to vote against climate deniers. If those kinds of candidates lose—especially in primaries and in state or local elections—it's going to be a lot easier for better candidates to run. And it's just as important to vote for people who promise real action. If they don't follow through, write to them, call them, or bring it up on social media—let them know they'll lose your support if they don't act on climate issues.

The United States has wasted way too much time and energy arguing over whether climate change exists, and the lack of US action has slowed international agreements on solving the problem. On the positive side, most governments around the world understand the seriousness of the situation and have pledged to cut emissions, even if many of them aren't yet on track to hit the agreed-upon targets. And the replacement of the science-denying Trump administration with president Joe Biden in the 2020 election suggests that the US government will likely return to a position of global leadership in fighting the crisis.

Biden nominated longtime climate advocates to several key positions and created a new cabinet-level position of climate envoy, filled by former senator and 2004 presidential candidate John Kerry. In his first day on the job in January 2021, Biden announced that the United States would rejoin the Paris Climate Agreement (of which it was a member until Trump withdrew) and took other climate-change action during his first week. The question is how much the new administration and Congress will try to accomplish, whether they succeed, and whether they can make up for lost time.

Government solutions to environmental problems can work—we've seen this plenty of times in the past. The element lead was widely used in the United States for a long time, in everything from paint to gasoline, even though it was highly toxic. Polluting businesses used to dump raw sewage, mercury, and other hazardous waste in the country's water on a massive scale without consequences. Dangerous materials like asbestos used to be common in building construction. All that changed because of federal laws and regulations. International agreements matter too, as we've seen on issues ranging from protecting endangered species to saving the ozone layer (see page 156).

It's going to take a combination of individual and collective action if we're going to get climate change under control. All the options discussed in this chapter are important, and protecting the world requires people all over the world to do their part.

Hopefully, you now have a better understanding of what's causing the climate crisis, the many ways it's affecting our world, and some idea of what's required to limit the damage. As we've discussed, thanks to human-made climate change, the world is not only a lot different than it was before the Industrial Revolution, but a lot different than it was just a generation ago. The problem has been more than a century in the making, but young people are the ones who will live through the worst of the crisis. We've already seen young people raise their voices in support of fixing the problem. They've inherited a mess, and it's going to take a lot of work by a lot of people to clean it up. That isn't fair, and it's not their fault.

Members of the Sunrise Movement, a youth-led activist group urging governments to fight against climate change (see page 186), hold up a banner as part of a protest in Chicago in 2019. The "12 years" refers to some predictions that unless the world makes significant changes to its energy usage and emissions by 2030, we are on track for devastating climate change effects.

But there's a chance here to rise to the occasion. If young people—like many of you reading this—get educated and active, they can literally save the world.

SOURCE NOTES

8   Interview with author, 2008.

10   Justine Calma, "This Was the Decade Climate Change Slapped Us in the Face," The Verge, December 10, 2019, https://www.theverge .com/2019/12/10/21003596/climate-change-end-of-decade-2019 -temperature-storms-wildfires-effects-emissions/.

11   Dane Placko, "City of Chicago Permanently Closes Two Beaches in Rogers Park," Fox 32 Chicago, January 16, 2020, https://www .fox32chicago.com/news/city-of-chicago-permanently-closes-two -beaches-in-rogers-park/.

11   Gaia Vince, "How Scientists Are Coping with 'Ecological Grief,' " The Guardian, January 12, 2020, https://www.theguardian.com /science/2020/jan/12/how-scientists-are-coping-with -environmental-grief/.

13   Interview with author, 2008.

13   Interview with author, 2008.

14   David Hudson, "President Obama: 'No Nation Is Immune' to Climate Change," The White House Blog, September 23, 2014, https://obamawhitehouse.archives.gov/realitycheck/blog/2014 /09/23/president-obama-no-nation-immune-climate-change/.

17   Jonathan Watts, "Global Warming Should Be Called Global Heating, Says Key Scientist," The Guardian, December 13, 2008, https://www.theguardian.com/environment/2018/dec/13/global -heating-more-accurate-to-describe-risks-to-planet-says-key -scientist/.

29   Steve Graham, "John Tyndall (1820–1893)," NASA Earth Observatory, October 8, 1999, https://earthobservatory.nasa.gov /features/Tyndall/.

34   Spencer Weart, "The Public and Climate Change (Cont.— Since 1980)," Discovery of Global Warming, Center for History of Physics, Accessed March 23, 2020, https://history.aip.org/climate/public2 .htm.

34   Philip Shabecoff, "Global Warming Has Begun, Expert Tells Senate," New York Times, June 24, 1988, https://www.nytimes.com /1988/06/24/us/global-warming-has-begun-expert-tells-senate .html.

35   John Noble Wilford, "His Bold Statement Transforms the Debate on Greenhouse Effect," New York Times, August 23, 1988, https:// www.nytimes.com/1988/08/23/science/his-bold-statement -transforms-the-debate-on-greenhouse-effect.html.

37   John Cook et al., "Consensus on Consensus: A Synthesis of Consensus Estimates on Human-Caused Global Warming," Environmental Research Letters 11, no. 4 (April 2016), https:// iopscience.iop.org/article/10.1088/1748-9326/11/4/048002/.

42   Fred Pearce, "How the 'Climategate' Scandal Is Bogus and Based on Climate Sceptics' Lies," *The Guardian*, February 9, 2010, https://www.theguardian.com/environment/2010/feb/09/climategate-bogus-sceptics-lies/.

44   "Gore Makes Sustainable Investment His Business," *The Age,* November 14, 2005, https://www.theage.com.au/business/gore-makes-sustainable-investment-his-business-20051114-ge18ga.html.

57   The World staff, "'It Is a Form of Injustice,' Caribbean Island Nations Struggle Against Rising Seas," The World, September 25, 2019, https://www.pri.org/stories/2019-09-25/it-form-injustice-caribbean-island-nations-struggle-against-rising-seas/.

59   Kara M. Watson, Glenn R. Harwell, David S. Wallace, Toby L. Welborn, Victoria G. Stengel, and Jeremy S. McDowe, "Characterization of Peak Streamflows and Flood Inundation of Selected Areas in Southeastern Texas and Southwestern Louisiana from the August and September 2017 Flood Resulting from Hurricane Harvey," USGS, 2018, https://doi.org/10.3133/sir20185070.

64   Barbara Crossette, "Severe Water Crisis Ahead for Poorest Nations in Next 2 Decades," *New York Times*, August 10, 1995, https://www.nytimes.com/1995/08/10/world/severe-water-crisis-ahead-for-poorest-nations-in-next-2-decades.html.

65   Associated Press, "The California Drought is Officially Over, but Next Could Be 'Around the Corner,'" *The Guardian*, April 7, 2017, https://www.theguardian.com/us-news/2017/apr/07/california-drought-over-jerry-brown-future-climate-change/.

78   "Chennai Water Crisis: City's Reservoirs Run Dry," BBC News, June 18, 2019, https://www.bbc.com/news/world-asia-india-48672330/.

88   Enele S. Sopoaga, "Keynote Statement," November 2015, https://unfccc.int/files/meetings/paris_nov_2015/application/pdf/cop21cmp11_leaders_event_tuvalu.pdf.

105   Jon Henley, "Icelandic Memorial Warns Future: 'Only You Know If We Saved Glaciers,'" *The Guardian*, July 22, 2019, https://www.theguardian.com/environment/2019/jul/22/memorial-to-mark-icelandic-glacier-lost-to-climate-crisis/.

110   Larry Elliott, "Global Warming Could Create 'Greater Migratory Pressure from Africa: David Attenborough Uses IMF Speech to Warn of Human Consequences of Inaction on Climate Change," *The Guardian,* April 11, 2019, https://www.theguardian.com/world/2019/apr/11/expect-even-greater-migration-from-africa-says-attenborough-imf-global-warming/.

114   Sandra Diaz et al., "Summary for Policymakers of the Global Assessment Report on Biodiversity and Ecosystem Services of the Intergovernmental Science-Policy Platform," Intergovernmental Science-Policy Platform on Biodiversity and Ecosystem Services (IBPES), May 6, 2019, https://zenodo.org/record/3553579.

114   Diaz et al.

126   Craig Welch, "Half of All Species Are on the Move—and We're
       Feeling It," *National Geographic*, April 27, 2017, https://www
       .nationalgeographic.com/news/2017/04/climate-change-species
       -migration-disease/.

129   National Audubon Society, "Survival by Degrees: 389 Bird Species
       on the Brink," https://www.audubon.org/climate
       /survivalbydegrees#climate2-survival-search/.

130   Caitlin Wall, "Ban Ki-Moon Warns of the Coming Water Wars,"
       Foreign Policy, January 24, 2008, https://foreignpolicy.com
       /2008/01/24/ban-ki-moon-warns-of-the-coming-water-wars/.

133   Sebastien Malo, "Cool U.S. Cities Prepare as Future 'Havens' for
       Climate Migrants," Reuters, April 6, 2019, https://www.reuters
       .com/article/us-usa-climatechange-migration/cool-u-s-cities
       -prepare-as-future-havens-for-climate-migrants-idUSKCN1RI061.

137   Nina Lakhami, "'People are Dying': How the Climate Crisis Has
       Sparked an Exodus to the US," *The Guardian*, July 29, 2019, https://
       www.theguardian.com/global-development/2019/jul/29/guatemala
       -climate-crisis-migration-drought-famine/.

148   Peter Schwartz and Doug Randall, "An Abrupt Climate Change
       Scenario and Its Implications for United States National Security,"
       FEMA, https://training.fema.gov/hiedu/docs/crr/catastrophe%20
       readiness%20and%20response%20-%20appendix%202%20-%20
       abrupt%20climate%20change.pdf, p. 14.

148   David Emery, "Did Donald Trump Claim Global Warming Is a
       Hoax?" Snopes.com, September 28, 2016, https://www.snopes.com
       /fact-check/donald-trump-global-warming-hoax/.

148   The White House, *National Security Strategy* (Washington, DC:
       GPO, 2015), 12.

152   Emily Bloch, "7 Inspiring Greta Thunberg Quotes Perfect for Your
       Climate Strike Protest Sign," *Teen Vogue,* September 16, 2019,
       https://www.teenvogue.com/story/inspiring-greta-thunberg
       -quotes-climate-strike-protest.

160   J. Poore and T. Nemecek, "Reducing Food's Environmental Impacts
       through Producers and Consumers," *Science* 360, no. 6392 (June
       2018): 987–992. https://science.sciencemag.org/content/360/6392
       /987.

SELECTED BIBLIOGRAPHY

Anguiano, Dani. "California's Wildfire Hell: How 2020 Became the State's Worst Ever Fire Season." *The Guardian*. December 30, 2020, https://www .theguardian.com/us-news/2020/dec/30/california-wildfires-north -complex-record.

Barron, Laignee. "143 Million People Could Soon Be Displaced Because of Climate Change, World Bank Says." *Time*. March 20, 2018. https://time .com/5206716/world-bank-climate-change-internal-migration/.

Borenstein, Seth. "Global Warming is Shrinking Glaciers Faster than Thought." The Associated Press. April 8, 2019. https://apnews.com /89bdd96ba86a445b93a53df09db784b4.

Borunda, Alejandra. "The Last Five Years Were the Hottest Ever Recorded." *National Geographic*. February 6, 2019. https://www.nationalgeographic .com/environment/2019/02/2018-fourth-warmest-year-ever-noaa-nasa -reports/.

Burt, Christopher C. *Extreme Weather: Climate Change Edition*. New York: W.W. Norton and Company, 2007.

Carrington, Damian. "Extreme Global Weather is 'the Face of Climate Change,' Says Leading Scientist." *The Guardian*. July 27, 2018. https:// www.theguardian.com/environment/2018/jul/27/extreme-global-weather -climate-change-michael-mann/.

Cox, John D. *Climate Crash: Abrupt Climate Change and What It Means for Our Future*. Washington, DC: Joseph Henry Press, 2005.

Dasgupta, Susmita. "Risk of Sea-Level Rise: High Stakes for East Asia and Pacific Region Countries." World Bank. March 9, 2018. https://blogs .worldbank.org/eastasiapacific/risk-of-sea-level-rise-high-stakes-for-east -asia-pacific-region-countries/.

EarthTalk. "Why Global Warming Can Mean Harsher Winter Weather." *Scientific American*. February 25, 2009. https://www.scientificamerican .com/article/earthtalks-global-warming-harsher-winter/#.

"How Climate Change Can Fuel Wars." *The Economist*. May 23, 2019. https://www.economist.com/international/2019/05/23/how-climate -change-can-fuel-wars/.

Fleischer, Jeff. "Adapting on the Atoll." Alicia Patterson Reporter. Spring 2009.

Freedman, Andrew. "Earth Sizzles Through October as Another Month Ranks as the Warmest on Record." *The Washington Post*. November 5, 2019. https://www.washingtonpost.com/weather/2019/11/05/earth-sizzles -through-october-another-month-ranks-warmest-record/.

Gander, Kashmira. "Deer, Sparrows and Magpies Are All at 'Substantial Extinction Risk' Because They Can't Adapt to Global Warming Fast Enough." *Newsweek*. July 23, 2019. https://www.newsweek.com/deer -sparrows-magpies-are-all-substantial-extinction-risk-because-they-cant -adapt-global-1450206.

Geiling, Natasha. "The Enduring Climate Legacy of Mauna Loa." *Smithsonian*. July 20, 2016. https://www.smithsonianmag.com/science -nature/enduring-climate-legacy-mauna-loa-180959859/.

Glick, Daniel. "The Big Thaw." *National Geographic*. https://www .nationalgeographic.com/environment/global-warming/big-thaw/.

Goldberg, Susan. "Fast-Melting Arctic Ice Poses Many Threats—Not All What You'd Expect." *National Geographic*. September 2019. https://www .nationalgeographic.com/environment/2019/08/editor-arctic-climate -change-poses-many-unexpected-threats/.

Greshko, Michael. "What Are Mass Extinctions and What Causes Them?" *National Geographic*. https://www.nationalgeographic.com/science /prehistoric-world/mass-extinction/.

Griffiths, James. "What Happens When Parts of South Asia Become Unlivable? The Climate Crisis Is Already Displacing Millions." CNN. July 27, 2019. https://www.cnn.com/2019/07/17/asia/india-nepal-flooding -climate-refugees-intl-hnk/index.html.

Hannam, Peter. "Human-Caused Global Heating Breaks Clear from Nature, Studies Find." *The Sydney Morning Herald*. July 25, 2019. https://www .smh.com.au/environment/climate-change/human-caused-global -heating-breaks-clear-from-nature-studies-find-20190725-p52akw.html.

Hansen, James E. and Makiko Sato. "Earth's Climate History: Implications for Tomorrow." NASA Goddard Institute for Space Studies. July 2011. https://www.giss.nasa.gov/research/briefs/2011_hansen_15 /PaleoImplications.pdf.

————. "Regional Climate Change the National Responsibilities." *Environmental Research Letters* 11, no. 3 (March 2, 2016). https://iopscience.iop.org/article/10.1088/1748-9326/11/3/034009/.

Harvey, Chelsea. "CO2 Emissions Reached an All-Time High in 2018." *Scientific American*. December 6, 2018. https://www.scientificamerican.com/article/co2-emissions-reached-an-all-time-high-in-2018/.

Harvey, Fiona. "Climate Change Soon to Cause Movement of 140m People, World Bank Warns." *The Guardian*. March 19, 2018. https://www.theguardian.com/environment/2018/mar/19/climate-change-soon-to-cause-mass-movement-world-bank-warns/.

Heos, Bridget. *It's Getting Hot in Here: The Past, Present, and Future of Climate Change*. Boston: Houghton Mifflin Harcourt, 2016.

Hersher, Rebecca. "2020 May Be the Hottest Year on Record. Here's the Damage It Did." NPR. December 18, 2020. https://www.npr.org/2020/12/18/943219856/2020-may-be-the-hottest-year-on-record-heres-the-damage-it-did/.

Holthaus, Eric. "James Hansen's Legacy: Scientists Reflect on Climate Change in 1988, 2018, and 2048." Grist. June 22, 2018. https://grist.org/article/james-hansens-legacy-scientists-reflect-on-climate-change-in-1988-2018-and-2048/.

Lynas, Mark. *High Tide: The Truth About Our Climate Crisis*. New York: Picador, 2004.

McKibben, Bill. *Fight Global Warming Now: The Handbook for Taking Action in Your Community*. New York: St. Martin's, 2007.

"More Near-Record Warm Years Are Likely on the Horizon." National Oceanic and Atmospheric Administration. Accessed January 2020. https://www.ncei.noaa.gov/news/projected-ranks/.

Morello, Lauren. "Arctic Ice Caps May Be More Prone to Melt." *Scientific American*. June 22, 2012. https://www.scientificamerican.com/article/arctic-ice-caps-may-be-more-prone-melt/.

Pierre-Louis, Kendra. "Heat Waves in the Age of Climate Change: Longer, More Frequent and More Dangerous." *The New York Times*. July 18, 2019. https://www.nytimes.com/2019/07/18/climate/heatwave-climate-change.html.

Porter, Eduardo and Karl Russell. "Migrants Are on the Rise Around the World, And Myths About Them Are Shaping Attitudes." *The New York Times.* June 23, 2018. https://www.nytimes.com/interactive/2018/06/20 /business/economy/immigration-economic-impact.html.

"Scientific Consensus: Earth's Climate is Warming." NASA. Accessed March 23, 2021. https://climate.nasa.gov/scientific-consensus/.

Sneed, Annie. "How is Worldwide Sea Level Rise Driven by Melting Arctic Ice?" *Scientific American.* June 5, 2017. https://www.scientificamerican .com/article/how-is-worldwide-sea-level-rise-driven-by-melting-arctic -ice/.

"Sources of Greenhouse Gas Emissions." US Environmental Protection Agency. Accessed March 23, 2021. https://www.epa.gov/ghgemissions /sources-greenhouse-gas-emissions/.

Stevens, William K. *The Change in the Weather: People, Weather, and the Science of Climate,* New York: Delacorte Press, 1999.

Tans, Pieter and Kirk Thoning. "How We Measure Background CO2 Levels on Mauna Loa." National Oceanic and Atmospheric Administration . Last updated March 2018. https://www.esrl.noaa.gov/gmd/ccgg/about /co2_measurements.html.

Waldman, Scott. "Climate Change Has Already Harmed Almost Half of All Mammals." *Scientific American.* February 15, 2017. https://www .scientificamerican.com/article/climate-change-has-already-harmed -almost-half-of-all-mammals/.

Weart, Spencer R. *The Discovery of Global Warming: Revised and Expanded.* Cambridge, MA: Harvard University Press, 2008.

Welch, Craig. "To Curb Climate Change, We Have to Suck Carbon from the Sky. But How?" *National Geographic.* January 17, 2019. https://www .nationalgeographic.com/environment/2019/01/carbon-capture-trees -atmosphere-climate-change/.

Wilkinson, Katharine. "The Woman Who Discovered the Cause of Global Warming Was Long Overlooked. Her Story is a Reminder to Champion All Women Leading on Climate." *Time.* July 17, 2017. https://time.com /5626806/eunice-foote-women-climate-science/.

Winston, Andrew. "The Story of Sustainability in 2018: 'We Have About 12 Years Left.'" *Harvard Business Review.* December 27, 2018. https://hbr .org/2018/12/the-story-of-sustainability-in-2018-we-have-about-12-years -left.

FURTHER INFORMATION

350.org
https://350.org
This international grassroots movement seeks to end the use of fossil fuels across the globe. Volunteers of all ages can join local 350 groups or start their own to take action. 350's website also provides up-to-date science on the link between fossil fuels and climate change.

AccuWeather Climate Change Center
https://www.accuweather.com/en/weather-blogs/climatechange
Check out the latest developments in climate science and weather changes from around the world on this news feed.

Asia Pacific Adaptation Network
http://www.asiapacificadapt.net
The Asia Pacific Adaptation Network is a platform for people, primarily those living in Asia and the Pacific region of the world, to share resources and knowledge about how to best prepare for and adapt to climate change. The site hosts a library of articles on technologies and ongoing projects as well as live chat and community events.

Center for Climate and Energy Solutions
https://www.c2es.org
This nonprofit organization focuses on developing practical solutions to climate change. The center worked on the famous Paris Climate Agreement and continues to work with experts to write and advocate for international climate change policy.

Citizens' Climate Lobby
https://citizensclimatelobby.org
The Citizens' Climate Lobby seeks to bring together people who believe in climate change and want to do something about it, no matter their political beliefs. The organization also trains volunteers on how to speak with elected officials about important laws or policies on climate change.

Climate.gov (managed by the National Oceanic and Atmospheric Administration)
https://www.climate.gov
This website provides the latest climate science news and data. Check out the Global Climate Dashboard to see how things such as temperature, snow melt, and glaciers are changing over time.

Climate Action Network International
http://www.climatenetwork.org
The Climate Action Network consists of member organizations from all over the world. View the members list to find organizations fighting climate change in your area.

Climate Reality Project
https://www.climaterealityproject.org
Climate Reality Project is a diverse international activist organization that trains people from all backgrounds to help fight climate change. The website provides "What You Can Do" links for tools and information on taking action now.

Climate Central
https://www.climatecentral.org
This organization of scientists and reporters publishes the latest on climate science with the goal of spreading awareness and informing the public of key findings.

Energy Star
https://www.energystar.gov
Check out Energy Star's website to identify the most energy-efficient appliances and find other ways to save energy in your home.

Environmental Defense Fund
https://www.edf.org
Having successfully gotten DDT, a toxic chemical, banned in the US as a result of its advocacy, the Environmental Defense Fund continues tackling the most pressing and urgent issues affecting the environment.

Friends of the Earth
https://www.foei.org
Friends of the Earth campaigns for environmental protection across the globe. The group challenges global leaders to do better by advocating for sustainable and socially just solutions to the climate crisis.

Greenpeace
https://www.greenpeace.org/global
Greenpeace uses direct action, peaceful protesting, and other forms of activism and advocacy to fight directly against environmental destruction and policies that would harm the environment.

Meatless Monday Campaign
https://www.mondaycampaigns.org/meatless-monday
Read more about Meatless Mondays and how reducing meat consumption
can help fight climate change.

National Aeronautics and Space Administration (NASA)
https://climate.nasa.gov/resources/education/
You might think of NASA as the space organization, but it also studies the
climate and publishes key findings in climate science. This page provides
lots of resources for those wanting to learn more about the science of climate
change.

National Geographic
https://www.nationalgeographic.com/environment/climate-change
Read National Geographic's latest articles on climate change and the
environment to get the facts.

National Resources Defense Council
https://www.nrdc.org
The NRDC works to protect all people's rights to a clean environment and
fights against polluters.

Nature
https://www.nature.com/nclimate/
This magazine is one of the most revered for its scientific publications. Read
the latest articles from climate researchers and scientists.

Plan International
https://plan-international.org/youth-activism/climate-change-activists/
Plan International educates youth on and advocates for their rights as a part
of the fight against climate change. The organization puts girls and young
women at the center of their fight, recognizing that, due to global gender
inequality, girls are disproportionately affected by climate change. The
website offers resources for young people who are ready to take a stand in
defense of their future.

Scientific American
https://www.scientificamerican.com/climate/
Scientific American publishes some of the latest research on climate change
and the environment.

Sunrise Movement
https://www.sunrisemovement.org
The Sunrise Movement is made up of youth activists who refuse to inherit a world ravaged by climate change. Learn about their important work and join the cause by visiting the website.

Union of Concerned Scientists
https://www.ucsusa.org
The Union of Concerned Scientists consists of about 250 scientists who believe that evidence-based solutions to climate change are necessary. Areas of their research include sustainable food and power, racial equity, and fighting back against misinformation.

US National Climate Assessment
https://nca2018.globalchange.gov
The National Climate Assessment is a report that gets updated every four years as required by the Global Change Research Act of 1990 and is delivered to Congress and the president of the United States. It assesses the climate crisis and its effects on all fronts, from agriculture to infrastructure to human health. You can read the 2018 report on this website.

World Resources Institute
https://www.wri.org
This international organization seeks to change the way we use Earth's natural resources so that our practices are sustainable and limit pollution.

World Wildlife Fund
https://www.worldwildlife.org
A major and renowned conservation organization, the World Wildlife Fund uses its resources to protect endangered species and at-risk environments.

INDEX

PHOTO ACKNOWLEDGMENTS

ABOUT THE AUTHOR

Jeff Fleischer is a Chicago-based author, journalist and editor. In addition to *A Hot Mess*, his books include *Votes of Confidence: A Young Person's Guide to American Elections*, *Rockin' the Boat: 50 Iconic Revolutionaries from Joan of Arc to Malcolm X*, and *The Latest Craze*. His fiction has appeared in more than sixty publications including the *Chicago Tribune's Printers Row Journal*, *Shenandoah*, the *Saturday Evening Post* and *So It Goes* by the Kurt Vonnegut Museum and Library. His journalism work has appeared in dozens of publications including *Mother Jones*, the *Sydney Morning Herald*, Mental Floss, *National Geographic Traveler*, and *Chicago* magazine.